The Archaeology of Medieval Ships and Harbours in Northern Europe

Papers based on those presented to an International Symposium on Boat and Ship Archaeology at Bremerhaven in 1979

edited by

Sean McGrail

National Maritime Museum, Greenwich,
Archaeological Series No. 5

B A R International Series 66

1979

B.A.R.

B.A.R., 122 Banbury Road, Oxford OX2 7BP, England

B. A. R. International Series 66, 1979: "The Archaeology of Medieval Ships and Harbours in Northern Europe".

ISBN 9780860540687 paperback
ISBN 9781407349800 e-book
DOI https://doi.org/10.30861/9780860540687

A catalogue record for this book is available from the British Library

This book is available at www.barpublishing.com

National Maritime Museum, Greenwich, Archaeological Series
General Editor—Sean McGrail, B. A. , Ph. D. , Master Mariner

1. Sources and Techniques in Boat Archaeology (B. A. R. - S29)
2. Logboats of England and Wales (B. A. R. 51)
3. The Graveney Boat (B. A. R. 53)
4. Dendrochronology in Europe (B. A. R. S51)
5. The Archaeology of Medieval Ships and Harbours in Northern Europe

CONTENTS

Page

LIST OF ILLUSTRATIONS

FOREWORD

From 9th to 11th February 1979 the second International Symposium of Boat and Ship Archaeology (ISBSA) was held at the Deutsches Schiffahrts-museum, Bremerhaven, Germany.

Members from Poland in the east to Wales in the west, from Norway in the north to Portugal in the south joined the symposium and decided to give themselves a very simple constitution to make sure that every third year a symposium will be arranged to give the opportunity to exchange the results of recent excavations or research.

For the third symposium in 1982 the members accepted the invitation of the Statens Sjöhistoriska Museum at Stockholm.

At Bremerhaven the surviving timbers of the Hanseatic Cog of Bremen had just been completely rebuilt. So this vessel served as a focus around which the papers of the symposium were arranged and these are now printed in this volume. I have to thank very much all contributors for their papers and Dr. Sean McGrail, National Maritime Museum, Greenwich, for the editorship, and I am grateful to Greenwich and to B. A. R. for publishing the papers in their series.

Dr. Detlev Ellmers
Director in charge
Deutsches Schiffahrtsmuseum, Bremerhaven

EDITOR'S INTRODUCTION

Of the 18 papers given at the Bremerhaven Symposium 14 are published here. Papers presented by A. Evans (Early Medieval Boat Finds), H. Roelfzema (Frisian cog-shaped boats) and V. Vogel (The Waterfront excavations at Schleswig and at Hedeby) will be published elsewhere.

Keywords have been listed at the end of each paper as an alternative to an Index. An attempt has been made to explain technical terms within the relevant texts or in illustrations. Where these are not clear reference may be made to the Glossary published in Sources and Techniques in Boat Archaeology (BAR Supplementary Series 29) or to that in Basil Greenhill's Archaeology of the Boat (A. & C. Black).

Information about the next meeting of the ISBSA may be obtained from the Archaeological Research Centre, National Maritime Museum, Greenwich.

I wish to thank Marilla Fletcher and April Whincop for their invaluable help in preparing this volume for publication.

Sean McGrail

LIST OF CONTRIBUTORS

Arne-Emil Christensen,
Universitetets Oldsaksamling, Oslo, Norway

Helen Clarke, B. A., Ph. D., F. S. A.,
Lecturer in Medieval Archaeology, University College, London

Ole Crumlin-Pedersen,
Viking Ship Museum, Roskilde, Denmark

Detlev Ellmers, Deutsches Schiffahrtsmuseum, 2850 Bremerhaven-M,
West Germany

Anthony W. Farrell, B. A.,
220, Wilson Avenue, Rochester, Kent

Octavio Lixa Filgueiras,
Professor of Architecture, Escola Superior de Belas Artes, Oporto, Portugal and Director, Department of Naval Ethnology, Centre for Historical Studies, University of Oporto, Oporto, Portugal

Thomas Gillmer,
Professor of Naval Architecture (retired), U. S. Naval Academy,
Annapolis, Maryland, U. S. A. 2140

J. de Jong,
IJsselmeerpolders Development Authority, Lelystad, Netherlands

Sean McGrail, B. A., Ph. D., Master Mariner,
Archaeological Research Centre, National Maritime Museum,
Greenwich, London SE10 9NF

Peter Marsden, F. S. A.,
Department of Urban Archaeology, Museum of London, 71, Basinghall Street,
London EC2P 2EX

Gustav Milne,
Department of Urban Archaeology, Museum of London, 71, Basinghall Street,
London EC2P 2EX

Reinder Reinders,
Ketelhaven Museum, 8251 PM Dronten, Netherlands

Eric Reith,
Musée de la Marine, Palais de Chaillot, 75116, Paris, France

Roy Switsur, M. A., Ph. D.,
Godwin Laboratory, University of Cambridge

Christiane Villain-Gandossi,
Centre National de la Recherche Scientifique (Paris), 8, Avenue du Château,
77410 Gressy-par-Claye, Souilly, France

1. THE COG OF BREMEN AND RELATED BOATS

Detlev Ellmers

Summary

The excellent condition of the Cog recovered at Bremen has allowed detailed examination of its construction. It is argued that this vessel is the most sophisticated example of the tradition to which cogs belong. By examining the constructional details and steering techniques of certain logboats, the 'julle'; the 'blockkahn'; the 'havelkahn' and the 'Ilmenau-Ever' boats, progressive features in the cog family's development are isolated and discussed. Finally, features of recent boats are shown to carry on this medieval boat-building tradition.

Something like a detective story was the way in which German and Danish scientists—historians and archaeologists—found out how the Cog, that big merchant vessel of the Hanseatic towns during the Middle Ages, had been constructed and how she had been developed from small coastal boats of the first millennium A.D.

The man who started to ask about the appearance of this type of ship was Paul Heinsius (1956). He was fortunate to find seals of Hanseatic towns with ships depicted on them which are referred to in written sources as Cogs. So Heinsius was able to define the shape of the cog, as a high-boarded ship in clinker construction with straight stem and stern-posts and one square rigged mast.

Though this definition covers just the side view above the waterline, the information sufficed for Siegfried Fliedner (1964) to identify the wreck of a ship, found during dredging operations in the River Weser some kilometres downstream from Bremen in 1962, as that of a Cog. The wreck was so well preserved, that Fliedner not only saw the bottom, as is common with most other ship finds, but also much of the sides above the waterline, especially both stem and stern posts. Had he not succeeded with his identification, nobody would have given any money for the excavation of a mere wreck. Being successful, he not only managed to get sufficient support for a complete excavation but also brought to light the key for the understanding of an hitherto unknown ship-building tradition.

The first man to use this key was Ole Crumlin-Pedersen, who in 1965 compared some boats and models of Danish ships with the Cog of Bremen and demonstrated their close relationship. They all had flat, flush-laid bottoms, steep clinker-built sides and straight stem and stern posts. This definition,

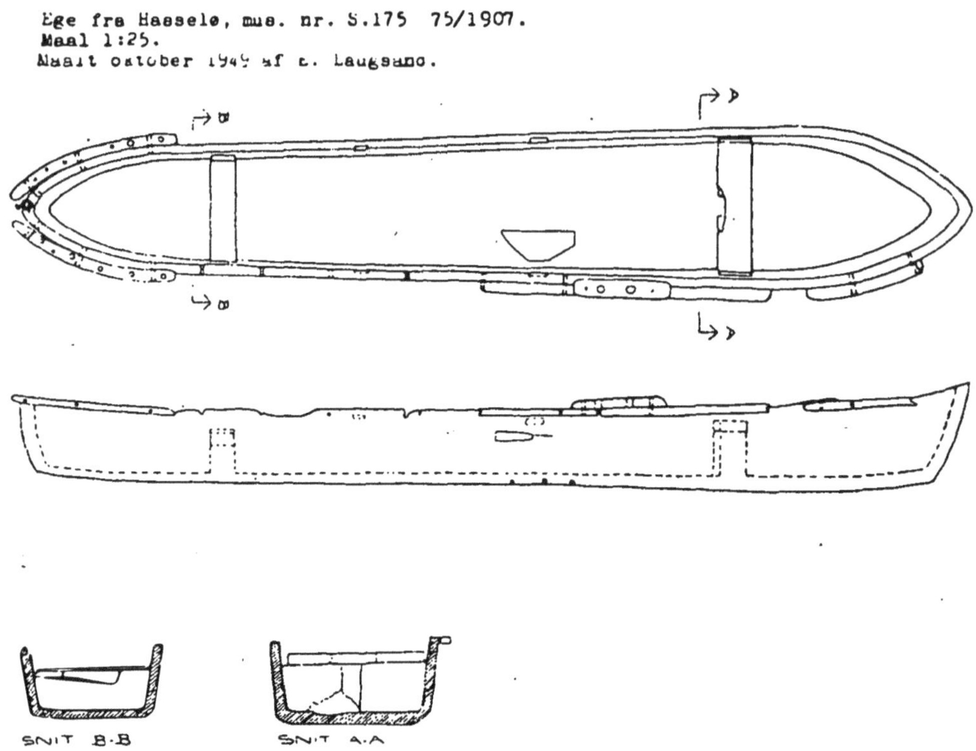

Fig. 1.1 A Cog-shaped logboat from Hasselø, Denmark, c. 1900 A.D. (after Rasmussen).

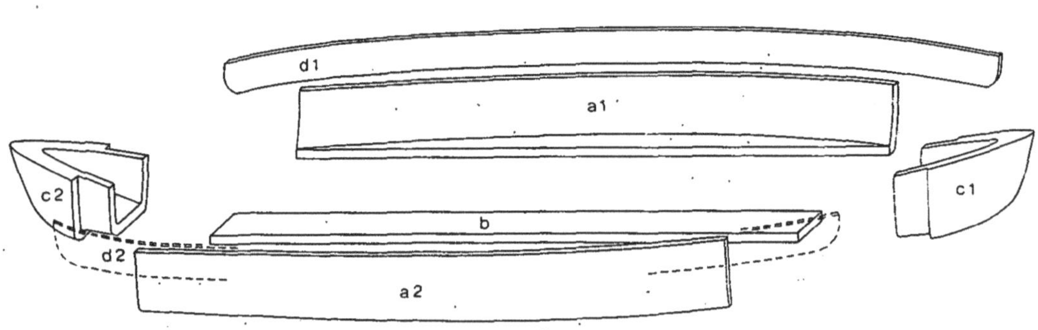

Fig. 1.2 Exploded drawing of a Block-Kahn from Gotmund near Lübeck, c. 1940 A.D.

which added a flat bottom to Heinsius's criteria, proved to be very helpful in deciding which wrecks were of Cog design and which were not, providing the hull survived.

When we started with the reassembling of the Cog of Bremen at the Deutsches Schiffahrtsmuseum in 1972 we not only studied her construction but also looked for boats fitting the above definition. We found quite a series of such boats, especially from Northern Germany, which in addition to matching this definition corresponded with the Bremen Cog in other details. The most important of these details is the fastening of the clinker planking by iron nails which had been hammered in from the outside and bent twice inside the boat so that their points penetrate the timber again. Small boats are caulked with moss only, with big vessels the moss caulking is kept in position by means of iron caulking clamps something like a butterfly in shape. This observation produced a further criterion for the ship-building tradition of the Cog; especially for archaeologists who may find little timber on a wreck site, but plenty of iron fastenings. They now are able to distinguish a Cog from a Viking ship even when all timber is gone, as the Cog-nails with their re-bent points differ very clearly from the rivets of a Viking ship with their diamond shaped roves. In this way the presence of Cogs in tenth century Birka and in all layers of Hamburg from ninth century onwards could be proved and the ship-fragments recently found in Denmark could be identified as those of Cogs.

Both the definition by shape and the definition by such constructional details as the nails not only served to decide whether a fragment of a hull belonged to the shipbuilding tradition of the Cog or to another, but also opened our eyes to understanding how the Cog herself had been developed from a simple craft into the most sophisticated ship of her whole family. However, this big cargo vessel of the late Middle Ages with her complicated construction was subsequently pushed into second place along the German shores in about 1400 A.D. by the Hulc which had been devloped into an effective vessel in English and French waters. About the middle of the fifteenth century the Hanseatic towns ceased to build Cogs and so the Cog of Bremen, having been built in 1380, marks the climax in the development of the type. In spite of the early end of the most developed member of this Cog family the smaller and more primitive boats of this tradition survived until the 19th or even 20th century. Thus, when establishing the sequence of typological development from primitive boats to the elaborated Cog, scientists not only depend on excavated boats or fragments of them but are able to draw their conclusions from boats still afloat on rivers and lakes and in coastal waters.

Having reached this state of knowledge we looked for the most primitive cog-like boat among excavated and recent boats and found a type of logboat which fitted the definition of the cog by shape. This boat had a flat bottom, steep sides (of course not in clinker technique as it was carved in one piece) and pointed ends which appear in the side view like straight stem and stern posts (Fig. 1.1). As the root end of the log served as the bow the boat was wider here than at the stern. Logboats of this type have been excavated in Schleswig-Holstein, but are not dated (La Baume, 1948). In some Danish islands this type survived till about 1900 A.D. (Rasmussen 1953). The interior of most of these boats was divided into three rooms by two bulkheads left there from the solid during the process of carving.

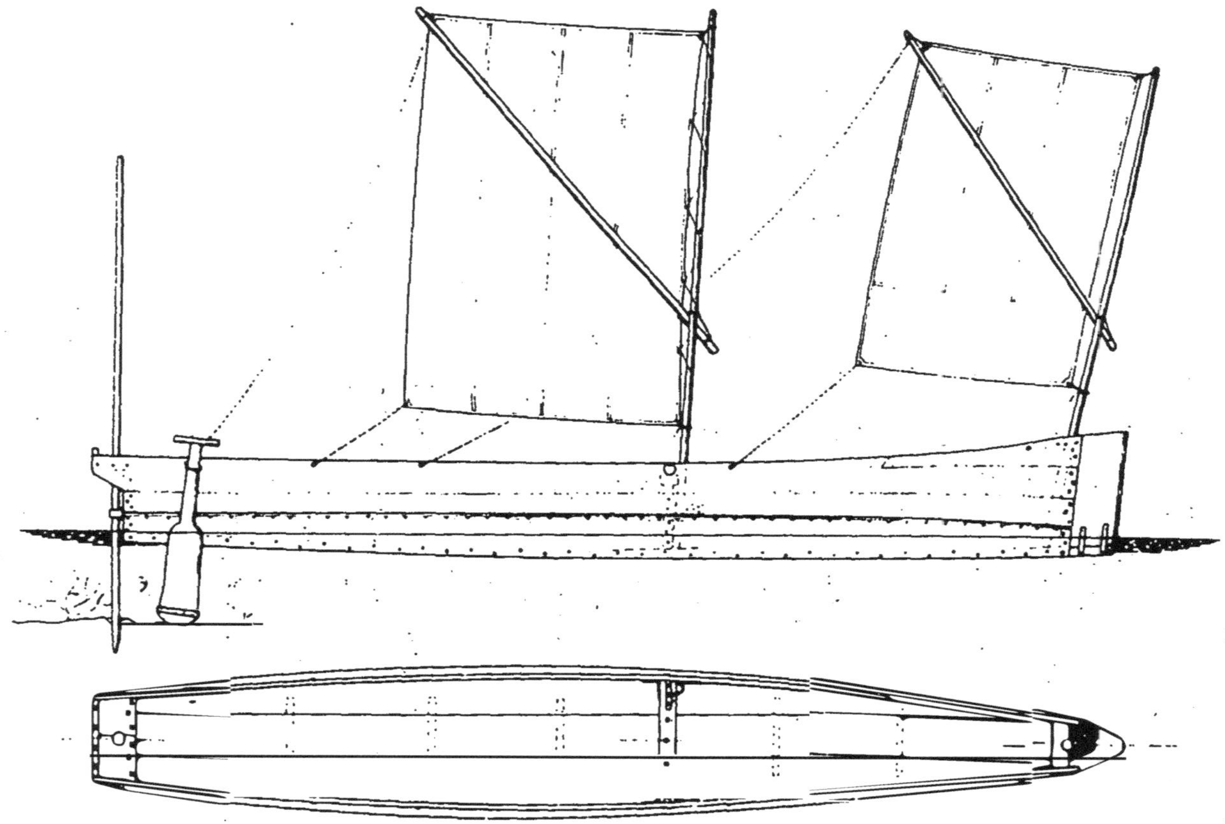

Fig. 1.3 Kahn from Steinhuder Meer near Hannover, c. 1930 A.D. (after Salemke).

When there was a lack of big trees people tried to build their boats in the shape of the old logboats but out of at least three single planks, one serving as the flat bottom shaped like a lancet with two pointed ends. The two others were nailed alongside the flat bottom at a distinct angle and met at both ends where stem and stern posts strengthened the construction. Knee-shaped ribs ran across the flat bottom with one arm supporting one or the other side alternatively. This type of construction is represented in the Deutsches Schiffahrtsmuseum by the 'Julle' from the Grosses Meer near Aurich, Ostfriesland, where it had been used for fowling until 1975.

Another unusual member of this family of boats is still used as a fishing boat in the Untertrave, downstream of Lübeck. To understand its construction one has to imagine what could be done with the logboat shown in Fig. 1.1 in order to increase its size: that is, how to get a boat broader than the tree trunk from which it was carved. The people near Lübeck split the log lengthwise, put one bottom plank in between and fastened these three parts together by nailing bottom timbers across them (Fig. 1.2). So they had a flush-laid bottom in shell-construction. But they were then faced with a difficult problem: at both ends there were gaps as wide as the added bottom plank which they had to close. They managed to do this by inserting the two ends of another logboat, the so-called "blocks" which gave the name of the type: "Block-kahn". These boats gave the wider boat the same shape as the logboat the people had started with. But in order to strengthen the construction and to get higher sides they fastened washstrakes in clinker technique to the top of the sides of the boat, using iron nails which they hammered in from the outside and bent twice on the inside in order to force the points into the timber again. Instead of ribs they used bulkheads which they inserted into the complete shell to make it stiffer.

In the Steinhuder Meer north of Hanover people used another variation of this construction as a boat for fishing as well as for peat and hay transport (Salemke, 1968). In trying to widen a logboat by the same method they found another way of filling the gaps at both ends. Instead of parts of another logboat they took a big solid log for the stem-post, triangular in cross-section, and at the aft end they inserted a broad transom (Fig. 1.3). The other constructional details are similar to those of the Block-Kahn: there are only differences in shape. As the Steinhuder Meer is a rather shallow lake the people had cut their logboats from the upper part of a horizontal log with the result that the sides taper to the inside, giving the boat a trapezoidal cross-section with the widest part at the bottom. They wanted to have the underwater part of the boat as wide as possible in order to get a shallow draft. An undated logboat of this shape has been found in that lake and it is clearly seen that the planked boat follows in all details the shape of this type of logboat with the sides still tapering inwards. Even the one bulkhead left in the solid across the middle of the logboat is copied in the planked boat by an inserted piece of timber, serving as the only transverse member in a boat that, in spite of nine metres length, has no ribs.

In the course of time people no longer used the half parts of logboats from the solid but learned how to construct this element from two separate planks, sawn from a log. Having reached this stage of construction they could close the gaps at the ends of a boat of the above mentioned type much

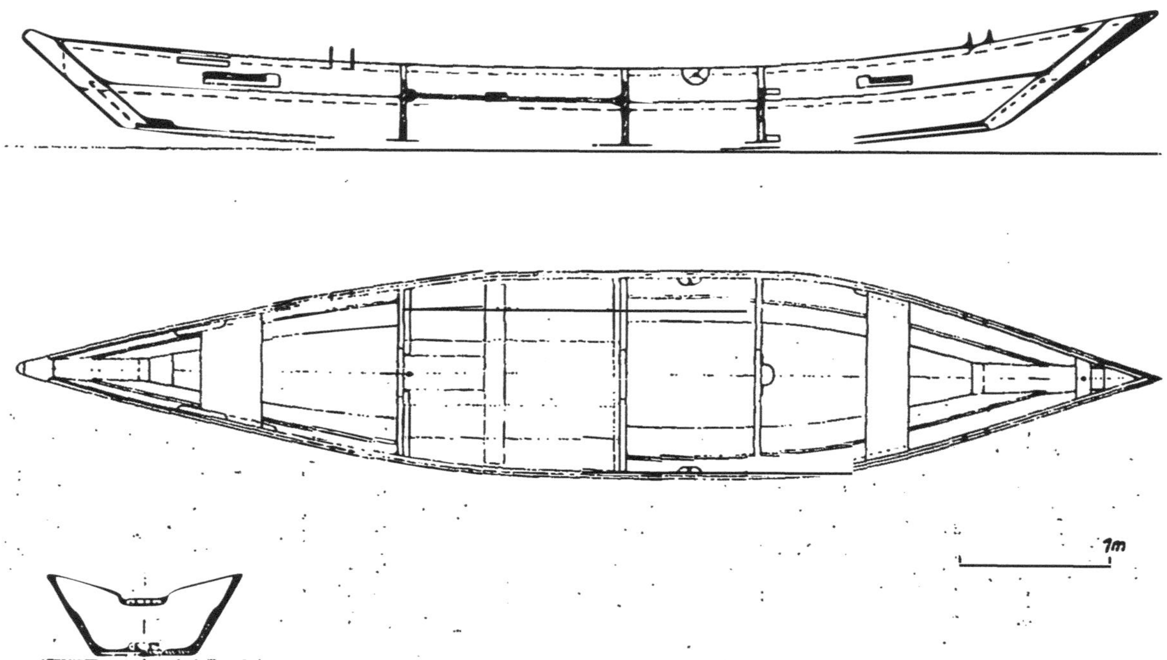

Fig. 1.4 Havelkahn from Potsdam, c. 1940 A.D. (after Rudolph).

a

b

Fig. 1.5 The Cog on Lübeck seals a) 1256 A.D., b) 1280 A.D. (after Ewe).

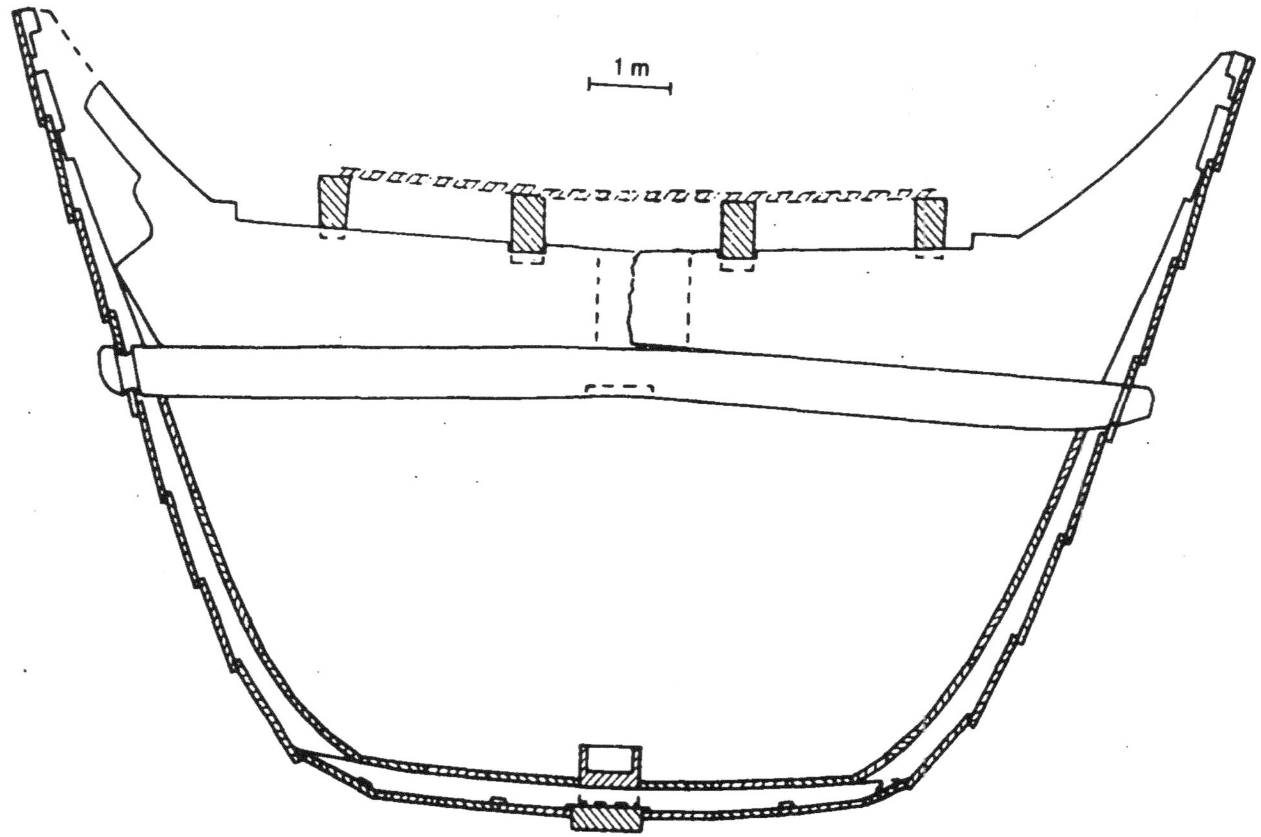

Fig. 1.6 Transverse section of the Bremen Cog, 1380 A.D.

more easily than by inserting "blocks" (fore and aft ends of dugouts) or other broad timbers: they extended the lowest side-planks to the ends of the boat where they met stem and stern posts. This state of development is represented in the Mueum by a fishing boat from the river Havel at Potsdam and Berlin built in 1940, (Fig. 1.4). This boat has two constructional details which cast light on some unusual features of the Cog of Bremen.

1. While the lowest strake fits into rabbets of stem and stern post the next strake overlaps both posts completely to meet the corresponding plank the other side. In the Cog of Bremen, the four bottom-planks fit into rabbets of stem and stern posts and the following eight side-strakes overlap both posts completely. But in contrast to the small inland boats the seagoing Cog has been made more seaworthy by covering this overlap with additional "false" stem and stern posts which run parallel to the inside ones, to which they are fastened by iron bolts with heavy heads protruding on the outside. These outside stem and stern posts are clearly to be seen on cog seals of the 14th century but are missing on the oldest ship-seals of Hanseatic towns such as that from Lübeck of 1256 (Fig. 1.5a). As this ship has a side rudder and animal heads on top of stem and stern posts and any sort of after-castle or the like is missing, earlier investigators thought that this ship was closely related to Viking ships and thus that Cogs had been developed from Viking ships or that there had at least been a strong influence from them. The foregoing analysis of Cog construction and related boats demonstrates that there is not any Viking influence to be seen. On the contrary, even the early Lübeck seal shows quite clearly that the strakes overlap both stem and stern posts. The depicted ship represents an early stage of development of the Cog, the outside posts having been introduced later in the 13th century: they are to be seen for the first time as narrow laths on the Lübeck seal of 1280 (Fig. 1.5b).

2. Another constructional detail of the "Havelkahn" of Berlin-Spandau are the four bulkheads which instead of ribs give internal strength to the hull. Though there are some boat-types with knee-shaped ribs belonging to the family of Cog-like craft, bulkheads from the very beginnings of their development played an essential role in the construction from the logboat to the Cog of Bremen. With her we find the bulkheads not down in the hold but on top of the cross-beams as a support for the deck-construction. The shape of these bulkheads is, in spite of their larger dimensions, very much the same as those from the "Havelkahn". This observation leads to the idea that the big seagoing Cog of late medieval date is something like two ships, one on top of the other with the crossbeams marking the transition (Fig. 1.6).

Indeed there is much to support this hypothesis. Not that people really put one ship on top of the other, but in order to increase the cargo-capacity they doubled the height of the side. They used their experiences with ships of half this height in so far as they repeated their methods when starting to construct the upper half of the side. Let us go into details of this: Mr. Roelfzema, Amsterdam, demonstrated during the symposium that in Dutch Friesland there are two different types of small Cog-shaped boat. Though they are built from sheet steel they exactly reflect the shape of wooden boats of the Cog family. Mr. Roelfzema had put one type of boat on top of the other: the result was a craft with a shape quite like the Cog of Bremen — admittedly on a smaller scale.

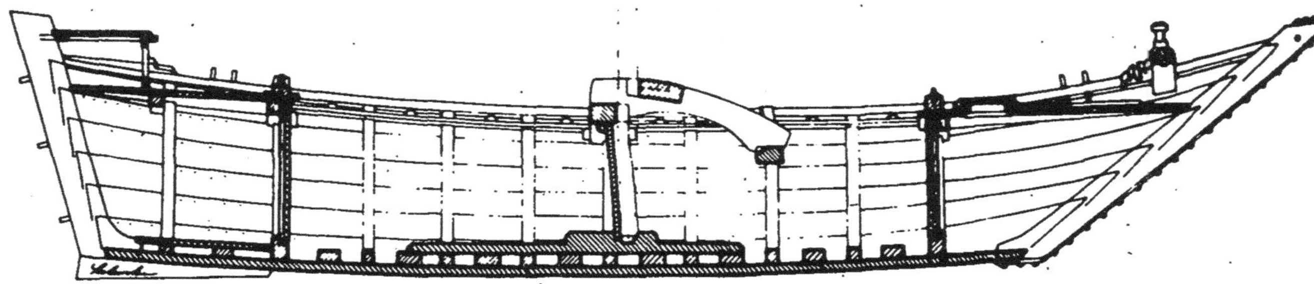

Fig. 1.7 Longitudinal section of an Ilmenau-Ever, 1710 A.D. (after Salemke).

The Bremen Cog herself tells us the story of how she was built after the same idea: people first laid out the strakes of the flat bottom, bent their ends to meet stem and stern posts and joined them together by means of bottom timbers. Next step was the clinker-construction of the four lowest side-strakes with their characteristic nails, the points of which are bent twice to enter the timber again. Then they installed the lower parts of the ribs, which sometimes cover the bent nails so we are sure that the ribs had been inserted into a pre-erected shell. On top of the fourth side-strake they put four of the big cross-beams — so to say, as bottom timbers of the "second ship" — and then continued with the next four side-strakes which they supported from the inside by the upper parts of the ribs and by four bulkheads on top of the four cross-beams. These bulkheads are by no means unique to the Bremen Cog; they had already been observed in wreck No. 5 from Kalmar by H. Akerlund (1951). This ship had also been constructed with the twice bent nails which Akerlund in his day did not know to be characteristic of a Cog.

To prove the hypothesis of a Cog being two ships one on top of the other we have to look for Cog-like vessels with sides half as high. And indeed they existed on North German rivers for much longer than the big Cogs, that is up to the end of the 19th century. A model built in 1710 of an "Ilmenau-Ever" from Lüneburg (Fig. 1.7) has been published by G. Salemke (1971) who saw the close relationship to the Cog of Bremen. What he did not observe, however, was the use of tiny iron nails bent twice even on the model. The flat bottomed hull is constructed with six side-strakes, each being narrower than the wide strakes of the Cog, so that the six side-strakes of the "Ever" form a side of roughly the same height as the four lowest side-strakes which form half the side of the Cog. Four cross-beams on top of the Ever-construction help us to understand that the cross-beams of the Cog are not only the beginning of a "second" ship but also the uppermost construction detail of the "first" ship to reinforce her frames.

We learn by this that the Bremen Cog shows in herself the history of the whole development of the construction of which she is the most developed example. The less developed boats, which survived the Cog for centuries help us to decipher the code of the constructional development not only of the Cog but of the whole boatbuilding tradition to which the Cog belongs. We see, for example, that all boats of this tradition — the Cog included — are a little bit drop-shaped in plan with the fore-end being definately wider than the aft end, as is natural for a logboat being excavated from a log with the root end fore and the crown-end aft. Not even the cog is outside this constructional schema.

In all Cog-type boats the flat bottom rises slightly at both ends, some times in a low curve, some times at a definite angle as with the Blockkahn near Lübeck, where the "blocks" are not in a horizontal line with the flat bottom but at a noticeable angle. Such an angle is to be seen on the Cog as well, where the middle bottom plank meets the knees to which stem and stern posts are fastened (Fig. 1.8). A similar feature is depicted on ships on some 9th century coins struck at Hedeby, and links them to the same boatbuilding tradition, (Fig. 1.9). These coins had been minted on the pattern of Carolingian coins of Dorestad in the Netherlands showing quite another type of

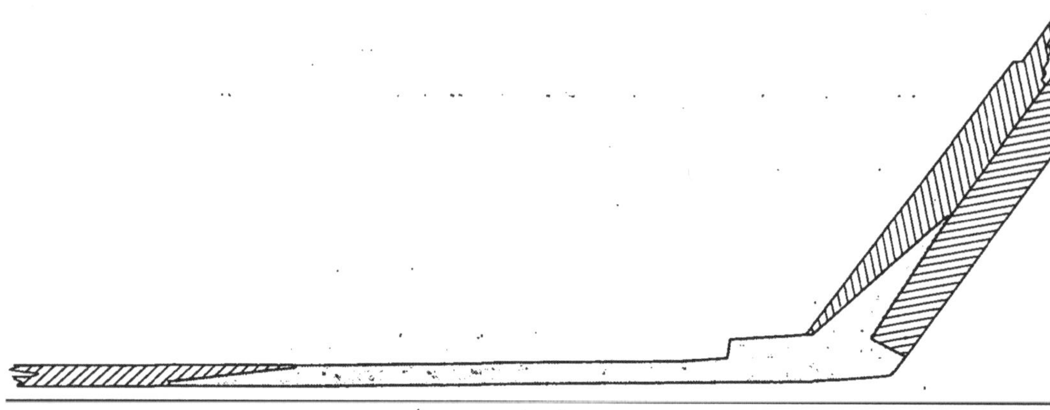

Fig. 1.8 Longitudinal section of the Bremen Cog, 1380 A.D.

Fig. 1.9 The Cog on Hedeby coins of 9th century A.D.

ship, the Hulc. Altering the type of ship on the coin means that the Hulc of the Dorestad coins played no role in the trade via Hedeby, where the Cog seems already in the 9th century to have been the dominating trading vessel (Ellmers 1972). Characteristic Cog nails from 9th century Hamburg and 10th century Birka confirm this observation, as well as the old name "kugg hamn" (= cog-harbour), for a certain harbour in Birka. Thus we are able to trace back the Cog far beyond the Hanseatic trade back to the time of Frisian trade with the Baltic.

The Cog-like boats of today not only help us to understand the construction of the big Cog but also help us de-decipher details in pictorial representations of early Cogs and so confirm that these recent boats indeed handed down to our time old stages of development. Most of the recent Cog-like boats are steered by ordinary stern rudders like the Cog herself, though the shape of these rudders vary very much, depending on the waters for which the boats are built (sea, shallow rivers, lakes or coastal waters). There is one boat, the 'Jülle', without any rudder at all, being poled in a shallow lake. The most interesting steering equipment is that of the Kahn of Steinhuder Meer (Fig. 1.3). The steering oar is called "Firrer" and looks like a paddle with a T-shaped handle and a broad blade. It cannot be turned but works like a leeboard by being put more or less deep into the water on the lee side of the vessel. The principle of operation is to have a fixed centre of effort in the sail somewhat aft of the middle of the boat and to alter the centre of lateral resistance of the vessel under water by lowering or raising the "Firrer". As the centre of effort in the sail is aft of midships the boat naturally turns into the wind. With the "Firrer" in a medium position, which varies with the direction and strength of the wind, the boat goes straight ahead.

The Cogs on the seals of Lübeck have been said to have used side-rudders of Viking type, as they are clearly shown at the side of the hull. But all ships with Viking type side-rudders on English seals (Ewe 1972) do not show T-shaped handles (as appear on seals of Lübeck and Neustadt near Lübeck) and the helmsmen hold tillers in their hands; whereas the Lübeck helmsman of 1280 has one hand on the T-shaped handle and the other on the shaft of the steering oar, demonstrating by this grip that there is no tiller at all (Fig. 1.5b). In other words: the steering oar of the early Lübeck Cogs was a "Firrer" like that of the Steinhuder Meer peatboats. It is shown on the three Lübeck seals in different positions: 1226 and 1256, the "Firrer" is kept in a resting position clear of the water, sloping down to the aft end of the vessel and protruding somewhat beyond it, so that it could be seen from the other side; the seal of 1280 shows the "Firrer" with part of the blade in the water. The seals of Neustadt near Lübeck and the seals of Niewport in Belgium show the "Firrer" held by the helmsman with the same grip as on Lübeck 1280, but the blade being completely out of the water. The Hedeby coins of the 9th century show the same steering equipment (Fig. 1.9): on all coins the main parts of the "Firrer' are hidden behind the ship, but some show it protruding beyond the stern of the Cog, that is in the resting position of the early Lübeck seals, while others show it protruding below the bottom of the Cog, at its deepest position to turn the ship out of wind. The use of these two positions (high or deep) demonstrates that these early Cogs did not use side-rudders of Viking type, but the "Firrer", which in those days was fairly widespread until it was succeeded by the stern-rudder on Cogs in the Baltic before 1240. Only in the small lake of Steinhude has it remained in use until today.

While some recent boats cast light on the development of the Hanseatic Cog by details which they preserved until today, there are other details, which do not appear on the Hanseatic Cog. There is, for example, a definite chine between flat bottom and sides on all boats. In spite of her flat bottom in the middle, the Cog has a smooth transition between bottom and sides and at both ends the horizontal bottom strakes are bent into an upright position to form pointed ends underwater for good sailing ability. Gentle curves of the underwater body are necessary for a seagoing vessel, which with a rectangular chine would roll sharply in the waves, whereas boats and ships of inland waters need not take heed of this fact and so can be built with a chine.

By comparing the Cog with boats of her shipbuilding tradition, we not only find remarkable details and developments, but we see that shipwrights building seagoing vessels in medieval times were far from being conserative, unlike their inland colleagues who built boats for lakes and rivers for hundreds of years in a pattern that they did not want to alter. Shipwrights in Hanseatic seaports used quite a lot of skill and imagination in developing their big seagoing vessels in size as well as in improvements of the construction, steering gear and so on. And we are able to follow these improvements step by step when using the analysis of recent boats as an additional method to get more information on the past (Ellmers 1976).

REFERENCES

Akerlund, H., 1951, Fartygsfynden i den forna hamnen i Kalmar (Uppsala, wreck V.

Crumlin-Pedersen, O., 1965, 'Cog—Kogge—Kaag', Handels-og Søfartsmuseet pa Kronborg Arbog, 81 - 144.

Crumlin-Pedersen, O., and others, 1976, 'Koggen med gullskatten', Skalk No. 6.

Ellmers, D., 1972, Frühmittelalterliche Handelsschiffahrt in Mittel- und Nordeuropa (Neumünster) 63-75.

Ellmers, D., 1976, Kogge, Kahn und Kunststoffboot (Bremerhaven).

Ewe, H., 1972, Schiffe auf Siegeln (Berlin, Bielefeld).

Fliedner, S., 1964, Die Bremer Kogge (Bremen). English Translation: Fliedner, S. and Pohl-Weber, R., The Cog of Bremen (1972).

Heinsius, P., 1956, Das Schiff der Hansischen Frühzeit (Weimar).

La Baume, W., 1948, Das Nydamschiff und die vorgeschichtlichen Einbäume in Schleswig-Holstein (Schleswig).

Rasmussen, H., 1953, 'Hasselø-Egen', Kuml 15-48.

Rudolph, W., 1966, Handbuch der volkstümlichen Boote im ostlichen Niederdeutschland (Berlin).

Salemke, G., 1968, 'Die aten Kähne vom Steinhuder Meer in Niedersachsen', Das Logbuch (Wiesbaden) 4, III: 3-6; 4, IV: 13-14.

Salemke, G., 1970-71, 'Die Ilmenau-Schiffe', Das Logbuch (Wiesbaden) 6, III: 29-30; IV: 3-8; 7, I: 7-14.

KEYWORDS

Medieval; 20th century.

Northern Europe; Baltic; Netherlands; Denmark; Bremerhaven; R. Weser.

Cog; blockkahn; havelkahn; Ilmenau-Ever; julle; logboats; seals; coins.

Ship and boat building techniques; typology.

2. DANISH COG-FINDS

Ole Crumlin-Pedersen

Summary

Research carried out in the past 20 years into the Cog ship type is summarised. The Kolding wreck discovered in 1943 and three wrecks recently excavated in Denmark are then described and identified as Cogs. This and other evidence is used to support the thesis that medieval sea-going ships were developed from a mixture of tradition, loans, and innovations.

The identification of the characteristics of the various types of medieval ships of Northern Europe known from the written sources has been a problem under discussion for several years. In 1956 the German scholar Paul Heinsius (1956) published his study of the ships of the early Hanseatic period in which he followed up previous attempts by other German scholars to identify and describe the cog-type on the basis of contemporary pictorial evidence and written sources. Mainly on the basis of the town seals he concluded that the characteristics of the cog were straight stem and stern posts, high sides and a deep keel. Heinsius rejected the idea of an interrelationship between the cog and earlier Frisian or later Danish local boat-types, and he claimed the cog-type to be a revolutionary invention, involving skeleton-first construction, introduced just before 1200 in the German area of the North Sea as well as in the Baltic.

This hypothesis was challenged in my article "Cog-Kogge-Kaag" (Crumlin-Pedersen, 1965). The objections to the Heinsius ideas were based on the archaeological evidence provided in the meantime by the Bremen cog-find and re-evalution of other finds, such as the Bruges boat (2nd century A.D.) and more recent parallels like the Nibe kaag-model from 1703.

My conclusions were that the characteristics of the cog-type were indeed the straight stem-posts and high sides, but that the straight bottom line represented a "carvel" (flush) keel-less bottom originating from the predecessor of the Hanseatic deep-sea cog, a Frisian coastal cog-type for tidal waters, and thus that no revolution had taken place, just an evolution.

This conclusion was adopted by Detlev Ellmers (1972) in his dissertation on the early medieval shipping of North Europe. Ellmers could contribute additional evidence in favour of the theory of the development of the cog from a flat-bottomed Frisian shiptype, though the final proof was still lacking.

In a recent follow-up on the cog-studies I drew up a set of principles in an attempt to explain the emergence of various medieval deep-sea ship types

(Crumlin-Pedersen, 1978). The theory is that a number of basically different approaches to the construction of a boat have been in use in various places, and that some of these well defined basic structures have existed in smaller vessels over very long periods. Under certain economic/technical/political conditions one or the other of these basic structures has served as the basis for development of a large shiptype dominating the scene for some centuries, until it is transformed under the influence of other types—or replaced by them.

According to this theory local small craft may in some cases serve as important clues to the understanding of the changing types of large sea-going ships we meet in the written sources and in the pictorial evidence, when we want to match the type-names with the wrecks found. Good examples of this (Fig. 2.1) are the Scandinavian Viking ships and the local boats of Western Norway today; another is the medieval cog and the recent Danish kaag; a third example is the medieval hulk (not yet definitely documented in the wreck-finds) and the ancient as well as recent boats built on the same principle; and the last example shown here is the punt-type, which during the Roman period grew to enormous size to suit transport-requirements on the Rhine.

For North European medieval shipwrecks this seems to be a useful hypothesis to apply to the ship-finds. Even those of which only the bottom-part has been preserved can be investigated according to this principle, because it is the bottom structure which in many cases enables us to trace the ancestry and type of the wreck.

With this in mind we will move to the Danish finds of wrecks, which could be characterized as cogs (Fig. 2.2). These are the finds at Kolding, Vejby and Kollerup. On the map is also marked the location of the Vigsø and Kyholm finds, which will be mentioned briefly in due course.

The first of these cogs to be found was the Kolding wreck. In 1943 a large side-rudder was found in the Kolding Fjord. This aroused a great interest in marine archaeology at the time. The National Museum, the Maritime Museum at Kronborg and the local museum joined forces with the Danish salvage company Switzer in an attempt to locate the wreck from which the rudder originated. In this they failed, but instead they found the wreck of a _c_. 18 m long vessel with an interesting structural layout, lying in mud at a depth of 4 m. The wreck was investigated by a professional diver as well as by Dr. P. Nørlund of the National Museum and Knud E. Hansen, naval architect of the Maritime Museum.

The sternpost showed rust-marks of fittings for a stern-rudder, and other pieces of wreckage recovered were beams, protruding through the sides (Fig. 2.3) as well as broad sideplanks cut to look like two overlapping clinker-planks on the outside. The mast-step was located well forward of amidships and a galley with hearth and pottery was found in the bow, as well as a coil of bast rope. A brief report (Hansen, 1944) was published in the Yearbook of the Kronborg Museum by Knud E. Hansen, who had taken careful notes of all recovered parts before they were buried in the mud again. Shortly afterwards the office of Knud E. Hansen was burnt out, and it was thought that all notes on the wreck had disappeared, until recently when a small black, partly burnt notebook was found with notes and sketches. This has given the clue to several details not explained in the published report; for instance a longitudinal stringer and timbers notched for this.

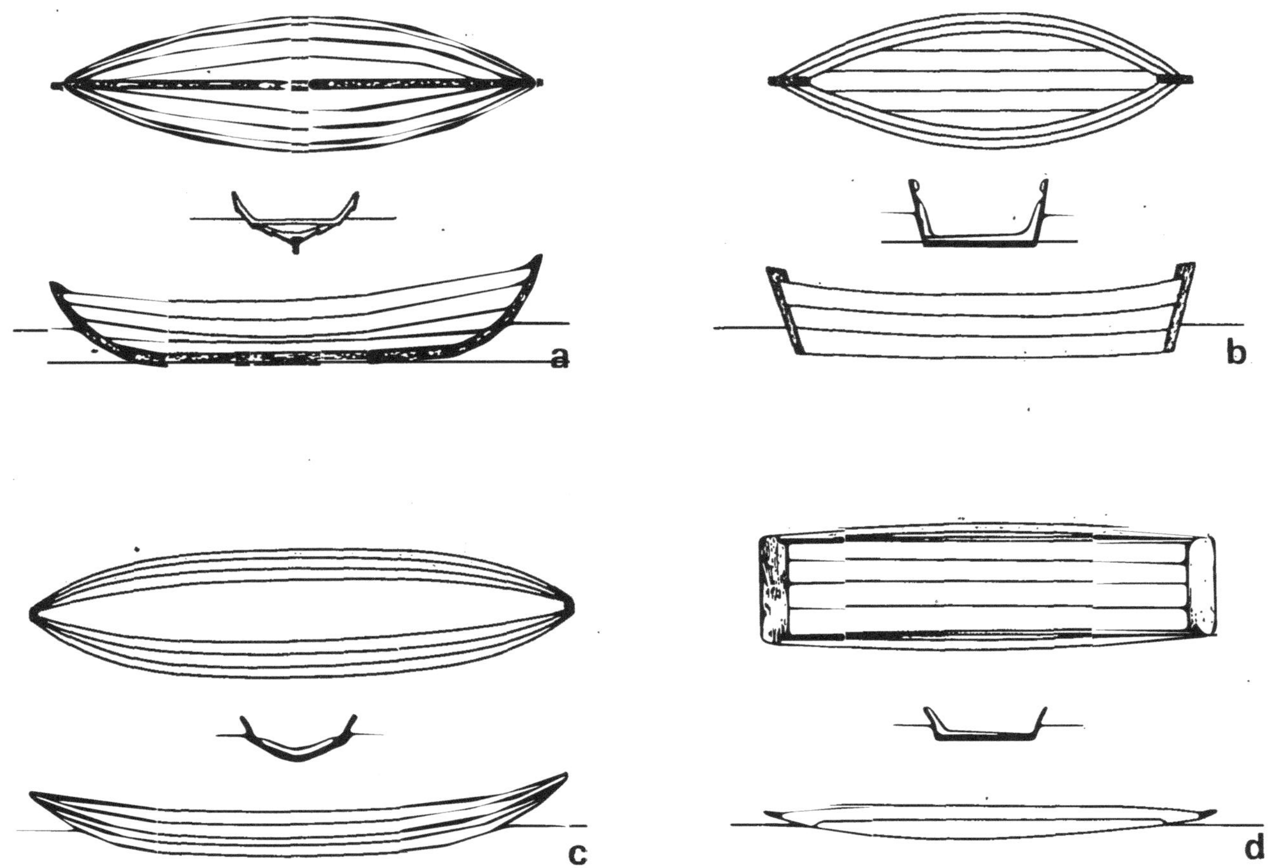

Fig. 2.1 Basic structural characteristics of four ship-types from early medieval Northern Europe: a) Nordic clinker-type, b) Cog- type, c) Hulk-type, d) Punt-type.

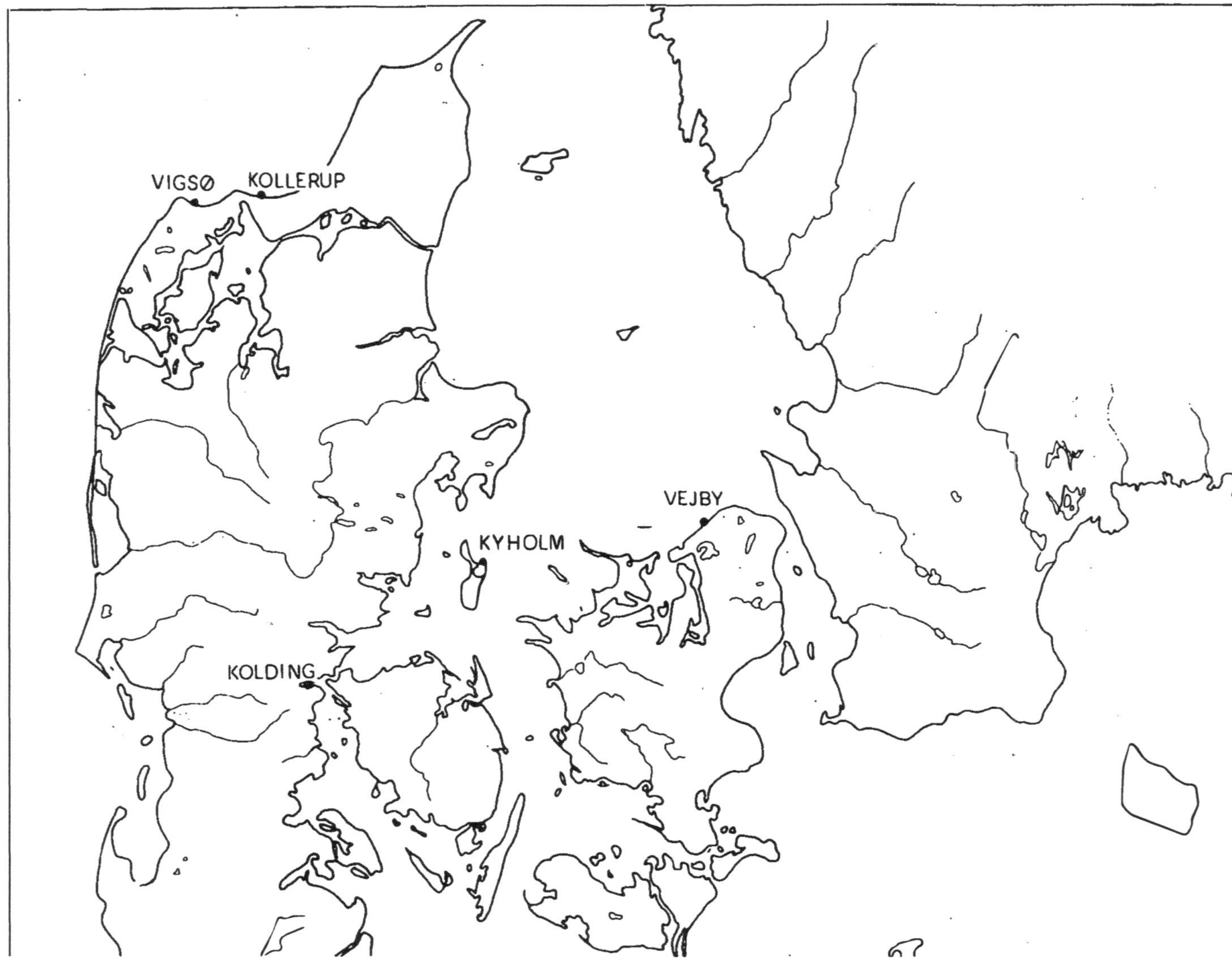

Fig. 2.2 Map showing the location of Danish ship-finds of Cogs and related types.

Fig. 2.3 Timber recovered in 1943 from the Kolding-cog. A beam notched to protrude through the planking.

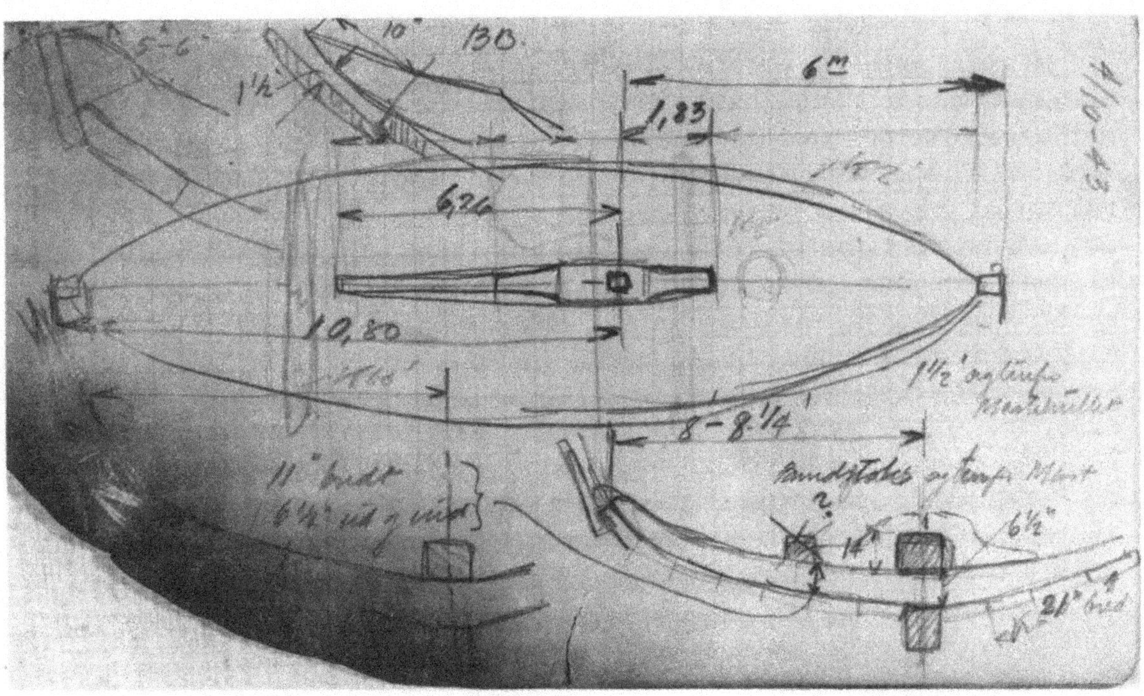

Fig. 2.4 Sketches of cross-section of the Kolding-cog preserved in the 1943 notebook of Knud E. Hansen, naval architect.

Of course there was no basis at the time to determine the shiptype, but it was especially noted that stem and stern were straight and that the hull was "carvel"-build in the bottom and "clinker" in the sides. The various sketches of the cross-section amidships (Fig. 2.4) show a keel, though no trace of this was observed inside the hull. As the divers did not see the underside of the hull this keel may well be disregarded, but there are rather good observations of the transition between the flush bottomplanks and the clinker-sides, and the position of the stringer is marked at the bilge.

There is no doubt on this basis that the wreck should be classified as a cog. Today the wreck is still lying in the deep mud, but there is a certain renewal of the old idea of raising the ship for exhibition in the local museum at Koldinghus together with other shipfinds from the vicinity, including the 11th-12th Cent. wreck from the drained Eltang vig, which was reported at the same time and excavated 1947 (Skov, 1952).

The Kolding fjord holds several other interesting medieval finds including a blockage and other boat finds, so this is definitely an example of a medieval maritime centre.

The next find is from a very different locality—from the weather-exposed Northwesterly corner of the Jylland-peninsular at Vigsø. Here sand and gravel is extracted in great quantities for construction purposes in Denmark and Germany and this material is taken up by suction in artificial lakes formed in the dune-covered beach area. Here two beautiful bronze figurines, probably early 15th century suddenly appeared in 1974 among the many cubic metres of stone and sand (Fig. 2.5).

Harald Langberg, curator of the medieval department of the Danish National Museum, immediately started a heroic fight to locate the wreck from which the figurines probably came. The idea was that a ship had wrecked and sunk on this spot with its cargo, and that the coastline had moved with the result that this wreck was now covered by 5-6 metres of sand and gravel, c. 500 m within the present coastline. To cut a long story short — we dived without success to locate possible parts of the wreck in the artifical lake, but by observing the material coming up at the continued suction on the site we found several lumps of rust, which turned out to be excellent moulds from which rubber casts could be made of iron objects from the wreck, mainly ship's nails, but also an axe, knives, locks and a set of rudder points (Fig. 2.6). This was clear proof of the fact that a vessel had actually wrecked on this very spot and had been partly destroyed by shipworms so that the iron nails and fittings had been loosened from the timber. The character of the iron objects corresponded exactly with those of the Bremer-cog and representations of cogs in town seals, and thus the Vigsø-wreck may well have been a cog. But we may never be able to get clear confirmation of this, as the wreck was not found during the subsequent work at the spot which included extensive drainage and excavation activities under extremely difficult conditions.

Fig. 2.5 Two bronze aquamaniles (early 15th century) from the Vigsø wreck-site.

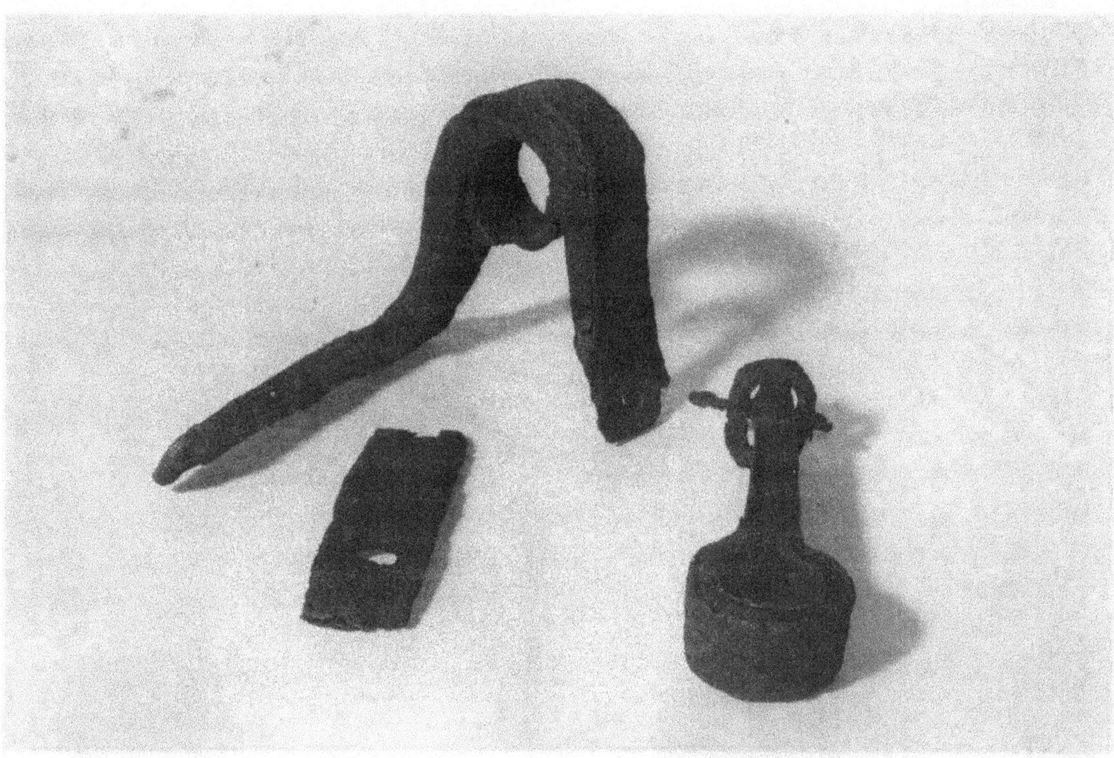

Fig. 2.6 Rudder pintles of the Vigsø wreck.

Fig. 2.7 1976 finds from the Vejby cog-site.

Fig. 2.8 Photo of midship part of the Vejby-cog after removal of the ballast stones.

In the summer of 1976 the scene changed to the North Sjaelland coast at Vejby, where high cliffs of clay make the coast most unhospitable for anything but small boats in fair weather. Here a couple of boys snorkled around an old wreck in 1-2 metres of water and found a pewter plate and gold coins lying around the wreck. The coins were 14th Century English nobles with fine representations of hulks on one side. The boys gave us notice and agreed to keep silent about the find until the finding-place had been thoroughly searched. With a day's notice we established a "picnic-camp" on the beach as a cover-up for our diving at the wreck to locate and record the remaining gold coins and other finds, and plotting these on a map of the site. This was all done so discreetely during ten days that none of the local residents realized that a treasure of 110 gold coins and several other objects lying around the wreck a few metres from the beach were taken in safe custody by a team of museum divers helped by the local boys (Fig. 2.7). A plan was drawn of the distribution of the objects showing a concentration of gold and other mixed metal finds around the stern. Only the last few days were spent on looking at the wreck itself, which was heavily loaded down with ballast stones. The stern part was cleared of stones and the cross-section was inspected and drawn showing that the three lower planks on each side of the keel-plank lay flush, while remaining planks were clinker. At the bow the ends of the strakes, the hood-ends, ended along a straight line, and the clinker planking was caulked with moss held by a flat batten on the inside—just as in the Bremen-cog.

With all these features there was little doubt that the wreck was a cog, wrecked on a voyage in ballast. As several other objects might lie around and in the wreck it was necessary to ban bathing and diving around the wreck, and yet it would be necessary to excavate the site fully to avoid further damage by the coastal erosion, which had exposed the finds and now threatened with further damage. Besides, the finding place of a hoard of gold coins is never safe against treasure hunters, so it was decided to try to find funds for a total excavation in 1977.

As headquarters for this work a former restaurant on the top of the cliff was excellent, and indeed we managed to find the £30,000 needed for the operation, which was carried out during five weeks in May-June 1977. A shore-based pump to cope with the sand and a diving vessel to take ballast stones ashore enabled us to uncover the wreck relatively quickly (Fig. 2.8) and to recover two coins from the Gdansk-region in the Baltic in the mast step. Halfway through we took one day off to discuss problems related to the find with colleagues from Danish and foreign museums as a most stimulating field academy session.

After having taken up 18 tons of ballast stones (originating somewhere along the Atlantic coast, probably Cornwall or Britany) the raising of the wreck could take place. It was complicated by the fact that there was not enough water for our diving ship to work over the wreck, and the exposed coast meant that we had to wait several days for calm weather. What we did was to build a large platform on which the wreck could be raised and transported to Copenhagen. This was floated on oil barrels to the site and on the day of the operation it was sunk in deeper water, the wreck was lifted by other barrels connected to a system of old firehoses, which had been washed

Fig. 2.9 The Vejby wreck on its platform in the tent-house after the raising.

Fig. 2.10 The Kollerup-wreck under excavation 1978.

Fig. 2.11 The mast frame of the Kollerup-wreck.

underneath the wreck, on to the platform, and the whole thing was raised by blowing air into the barrels. In principle this was easy enough, but we met with many problems due to the shallow water before the wreck was properly seated on the platform and well above water. By the end of the 16-hour working day the cog could finally be towed to the nearest harbour, where it was lifted on to a large truck for transport to Copenhagen with assistance from the police and fire brigade to keep the ship wet all the time.

A tent-house was prepared in advance to receive the wreck (Fig. 2.9), and here it was kept for some months while detailed drawings were prepared and water tanks were build to hold the timber. Here the details could also be studied more closely, as for example the lower part of the stem which compares strikingly with the Bremen cog. In the end the wreck was taken apart and the individual parts drawn full scale and packed down for future conservation and display at Roskilde.

On the basis of the associated material — the coins, metal objects and ballast stones — this vessel seems to be a cog build on the southeastern Baltic coast and wrecked around 1380 on a passage home from Western Europe (Crumlin-Pedersen, et al, 1976). On entering the Sound she seems to have come off her course in bad weather, tried to anchor but was driven on to the shore and broken to pieces by the surf. (A length of her anchor-cable was found underneath the ship, tight). Her master and crew may well have drowned as the gold remained untouched in the sea-bed in 2 metres of water for 600 years.

This was in 1977. But nowadays medieval wrecks tend to turn up in Danish waters each year, and 1978 was no exception to this rule. This time the scene was again the Northwest coast of Jylland, at Kollerup, and again the wreck was located by gravel suction several hundred metres within the present coastline. Conditions were very similar to those at Vigsø, where the figurines were found, except that here the wreck was definitely present, lying just below the water level. It was established from the few parts of the wreck exposed for study that the construction followed the well known cog pattern, though in this case the wreck was probably an older version than the Vejby cog. This might not be a new Vigsø-find with valuable cargo in the ship, but Jette Kjær, curator of the Thisted Museum, decided to try to raise funds for a complete excavation of the ship. This was indeed a brave decision which has provided this museum with one of the most important shipfinds of the 1970s — and with all the problems you meet when working with this sort of material.

The excavation took place in the autumn of 1978 with Hans Jeppesen as field director and with expert assistance from Arhus University (geology) and from Roskilde. The site was drained and the wreck was excavated (Fig. 2.10), showing that the ship had wrecked on a sand bank with both stem and stern at a lower level than the midship part. About 2 m of the straight stem and stern posts had survived the attacks which local people had made with axes on the wreck after the ship had stranded.

The details of this 20 m long wreck were of great interest. There was no keelson, but a maststep in a very large transverse bottom timber (Fig. 2.11).

In this same timber there was a notch for a transverse bulkhead aft of the mast. Another notch in the same timber was in the side facing aft at a level corresponding to the height of the midship-frames, indicating that the ship, which had no fixed ceiling, had had loose planks lying tightly on top of the frames to separate the cargo from the bilgewater. The bottom planks were flush/carvel for as much as six planks on either side of the centre-plank amidships, but at each end only two or three planks lay flush — the others changed over to overlapping/clinker-fashion. On one side a few of the upper planks were preserved with a stringer sitting directly on the outer planks and with an upper frame passing over the stringer.

From the notches cut in the top-ends of the floor-timbers it seems that this stringer ran along both sides over the greater part of the length of the vessel. Strangely enough there seems to have been little or no contact between the floor-timbers and the support for the upper part of the sides. And indeed the sides have had a certain height, as planks from a bulkhead aft have been preserved for a length of 2.5 m.

The craftsmanship is very good with the frames trimmed off very neatly, with slender stem and stern posts, and with a very careful mending of the many cracks in the timber, showing that this vessel was build at a place with a long tradition in shipbuilding, but a poor supply of timber — parallel to the Bremen cog — but in contrast to the Vejby cog.

The position of the mast is strikingly far forward in front of the hold, and the wreck is narrow in relation to its length. On the basis of pottery and other objects from the wreck this is preliminary dated to the 13th Cent. The wreck has now been taken apart and is stored in water-tanks, awaiting future conservation and presentation to the public.

This Danish cog-find may serve us as a good guide in the search of the origin of the cog-type. Vessels of similar mast-supporting structures have been found along the upper and lower Rhine, Bevaix (1st Cent.), Bruges (2nd Cent.) where we also find the mast situated well forward of amidships, and the same applies to the Blackfriars (2nd Cent.) ship from the Thames. It thus seems to indicate that the early Hanseatic cogs were modelled on principles, which have been called Celtic, and which were most likely developed by the Frisians into sea-going vessels. The Kollerup cog may well have been one of the first cogs to have tried to sail past the Skaw, as we hear about these Ummelands-farere for the first time by the middle of the 13th Century.

To sum up, we can now follow a tendency leading from what we may consider a rivercraft with extra sideplanks and a wide flush bottom into a structure with a fully intergrated frame structure and very few planks deviating from the clinker principle (Fig. 2.12). And when the Danish cog finds are shown to scale in longitudinal sections together with the Bremen cog, it is evident how the support and position of the mast changes to suit the tradition for a mast located centrally in a keelson, as in Scandinavian ships (Fig. 2.13).

But this was not a one way influence, which may be demonstrated by the last find to be mentioned here, the Kyholm-wreck. This was located in 1977 in a sheltered natural harbour off Samsø, and in 1978 we investigated the site to get further information on the wreck and coastal changes in the area. The

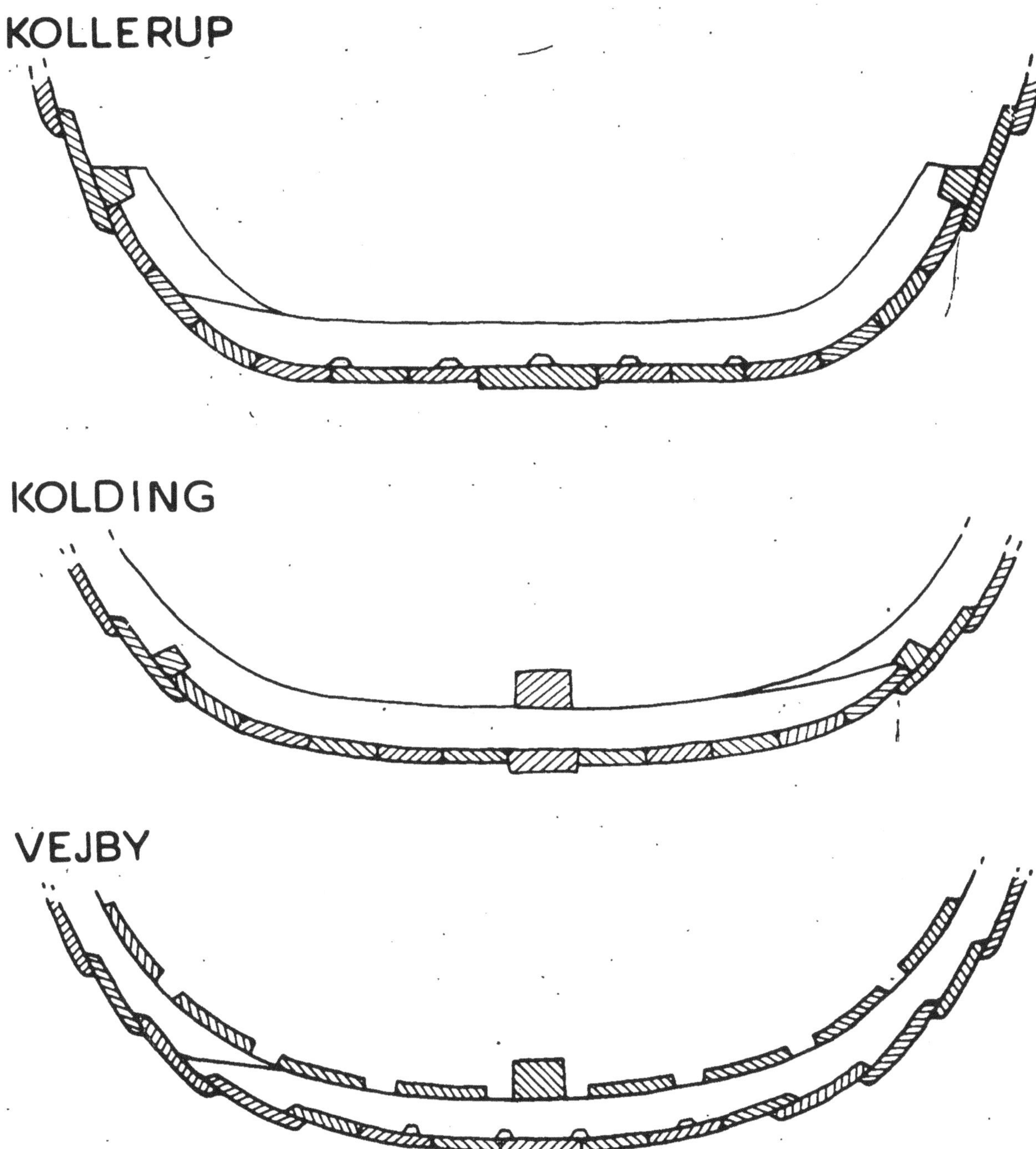

Fig. 2.12 Sketch of cross-sections of three Danish cog-finds.

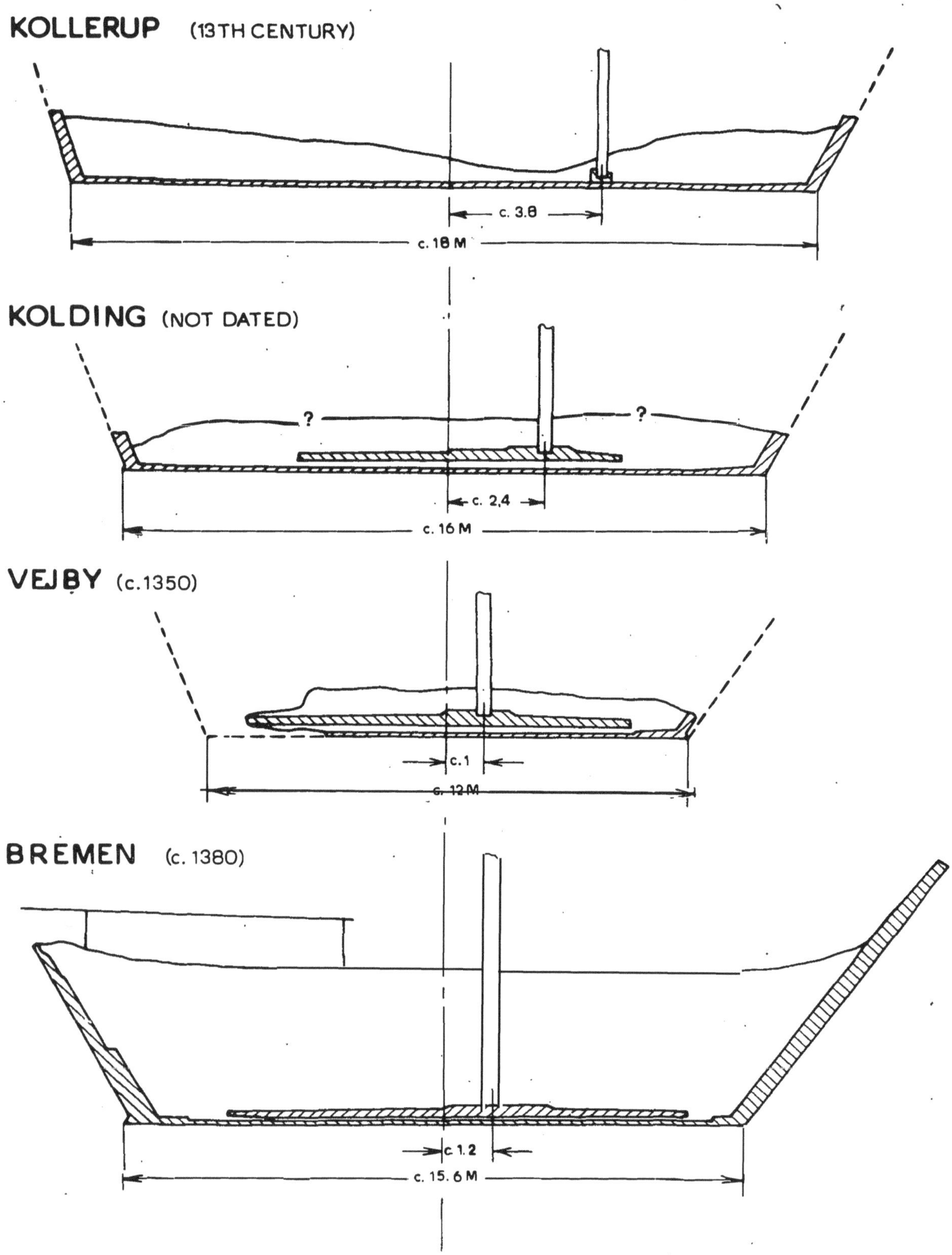

Fig. 2.13 Longitudinal sections of Cog-finds.

Fig. 2.14 The stem of the Kyholm-wreck *in situ*, 1978.

wreck was excavated, recorded and covered over again, which showed that it was clearly of the Scandinavian tradition, following the pattern of other Scandinavian ship-finds of the 12-14th Centuries.

As we first saw her stem from the inside it was a perfect parallel to the stepped Skuldelev-3-stem, but once it was fully excavated (Fig. 2.14) and taken up, we realized that the shipbuilder had had an ambition to follow the latest fashion and thus he had cut the forward edge straight to give the ship some of the outward appearance of the cog. I think we can get no better illustration of the fact that in the history of shipbuilding we have a bottom layer of local traditions of various basic structures and an upper layer producing sea-going ships out of a tantalizing mixture of tradition, loans and structural changes to meet new challenges.

REFERENCES

Crumlin-Pedersen, O., 1965, 'Cog-Kogge-Kaag. Træk af en frisisk skibstypes historie'. Handels- & Søfartsmuseet pa Kronborg, årbog 1965: 81-144.

Crumlin-Pedersen, O., 1978, 'The Ships of the Vikings'. The Vikings. Proceedings of the Symposium of the Faculty of Arts of Uppsala University June 6.-9. 1977. Uppsala: 32-41.

Crumlin-Pedersen, O., et al, 1976, 'Koggen med Guldskatten' Skalk, 6: 9-15.

Ellmers, D., 1972, Frühmittelalterliche Handelsschiffahrt in Mittel- und Nordeuropa. Schriften des Deutschen Schiffahrtsmuseum Bremerhaven. Bd. 3, Neumünster.

Hansen, K.E., 1944, Kolding Skibet. Foreøbig Meddelelse om Fund af Middelalderskib'. Handels- & Søfartsmusset paa Kronborg, Aarbog 1944: 119-129.

Heinsius, P., 1956, Das Schiff der hansischen Frühzeit. Quellen und Darstellungen zur hansischen Geschichte. Neue Folge, Bd. XII, Weimar.

Skov, S., 1952, 'Et middelalderligt skibsfund fra Eltang Vig'. Kuml, Årbog for jysk arkæologisk selskab 1952:65-83.

KEYWORDS

Medieval.

Denmark.

Cog; Kolding wreck; Kollerup ship; Vejby ship.

Shipbuilding characteristics; evolution of sea-going ship types; underwater excavations.

3. MEDIAEVAL SHIPS: RECENT FINDS IN THE NETHERLANDS

Reinder Reinders

Summary

Two areas in the Netherlands offer great possibilities for discoveries of shipwrecks: the IJsselmeerpolders — the seabed of the former Zuiderzee — and the area along the river Rhine. The vessels discussed in this paper are classified into three types: firstly ships made of logs, secondly flat-bottomed vessels, thirdly a fishing vessel, called Waterschip.

Besides all differences in shape, construction and function these vessels have one element in common: until the 16th century shipbuilders used moss, lath and iron clamps for caulking the seams.

INTRODUCTION

The map of ship finds in Ellmers' Mittelalterliche Handelsschiffahrt (1972) shows six dots in the area that is nowadays the Netherlands. Of this number three dots represent finds of Roman times and three represent Mediaeval ships. At first sight a disappointing number, but now seven years later in 1979 there are about twenty five, too many, I am afraid to discuss all of them.

The mediaeval ships were found in two areas where, due to elaborate works, many archaeological finds appear. One area is the former Zuiderzee. The Zuiderzee law enacted in 1918 included not only the Afsluitdijk, the barrier dam, but also the construction of five polders. Four of these polders have now been completed. When reclaiming the Wieringermeer in 1930, the archaeologists showed no interest in shipwrecks, because they were of too recent date. The investigations into shipwrecks started in 1942, during the reclamation of the Northeastern polder. Up till now 330 shipwrecks have been found (Fig. 3.1). A great number of them have been excavated, among them about twenty vessels of mediaevel date.

Another part of the Netherlands where many ships have been found due to building activities and sand dredging is the river Rhine area. During the last few years information about ships from Roman times has been increased by the finds from Zwammerdam, Druten, Kapel Avezaath and Woerden, besides the old find of Vechten. Ships of mediaeval date have been discovered in the town of Utrecht and near Meinerswijk, south of Arnhem.

I should like to discuss briefly three kinds of vessels found recently in these two areas.

- ships made of logs
- flat-bottomed vessels
- a fishing vessel, called: waterschip.

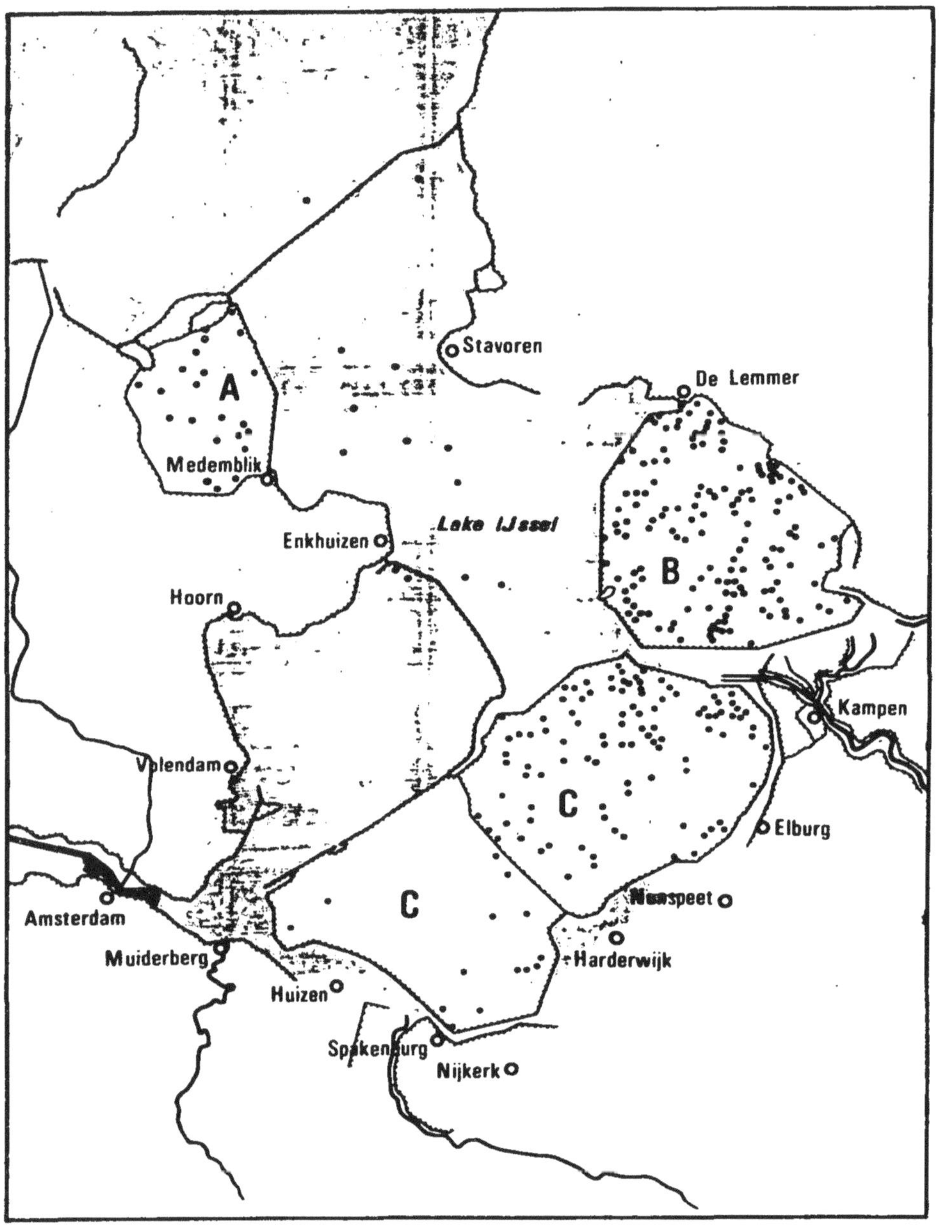

A Wieringermeer
B Northeasternpolder
C Flevoland

Fig. 3.1 Shipwrecks in the Lake IJssel area

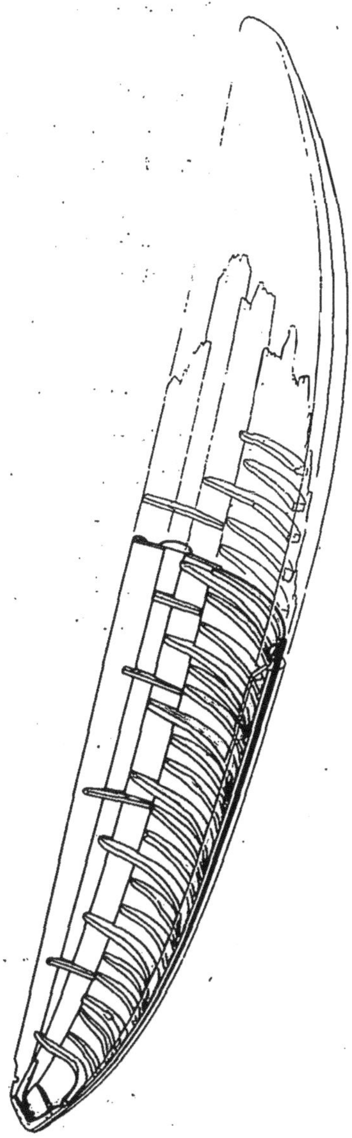

Fig. 3.2 Utrecht 2

SHIPS MADE OF LOGS

First there is the well-known Utrecht ship discovered in 1930 in a former bed of the Rhine, radiocarbon dated to the 8th century. In 1974 two other ships were discovered in Utrecht, one of them identical to the ship found in 1930, and dating from the 12th century (Hoekstra, 1975). The most striking element is the dug out and expanded log used as the bottom of the vessel, with an extension on each side of two planks connected by half a log (Fig. 3.2).

Unlike the first Utrecht ship (Ellmers, 1972:292), this one had some side frames.

Another ship from the Rhine area was found in 1976 near Meinerswijk. It was one of three ships found in a sandpit. A fragment of 6 m was recovered, the rest of the ship probably lies in the sandpit under water covered by sand. The vessel consisted of two equal halves of a log, split along its longitudinal axis. After hollowing, the two halves served as sides of the ship. Three planks were inserted between these elements and two were fixed at the top of the sides. Bottom and sides were fastened to the floor timbers and side frames by trenails (Fig. 3.3). The foreward end consisted of three planks with an elevation of 55 degrees; the after end could not be recovered.

I doubt whether the name oberländer for this type of vessel, as suggested by some archaeologists, will prove to be correct. In the Dutch language the words "overland-werk" and "bovenlander" were explained as: orginating from the upper Rhine area or vessel with overlapping strakes (Sopers, 1974:33). In any case the log and the moss of the Meinerswijk vessel came from a lowland area and the vessel had no overlapping strakes.

FLAT-BOTTOMED VESSELS

A shipwreck from lot Q 75 in the Northeastern polder is in many discussions considered as one of the earliest examples of the cog family. The date given for the wreckage of this ship, the 12th or even 11th century, is based on a layer of shells, usually dated to the 13th century, laying above the shipwreck. (Van der Heide, 1974:421.)

Re-examination of the soil survey showed that the layer of shells was not above but below the shipwreck. So consequently Q 75 should be dated to after the 13th century.

An early 14th century date corresponds very well with the date of another ship of the same construction, excavated in 1976 at lot N 5 in Flevoland. The bottom of this ship is flat, carvel built, the sides are clinker built (Fig. 3.4). The vessel has been been built on a keel; the fore and after part of the keel having been extended with a knee timber, the upright parts of the knee forming the base of stem and stern. Both fore and aft the carvel construction of the bottom passes into the clinker construction of the sides, as can also be seen with another ship found at lot K 73/74 (Fig. 3.5). Considering the length/beam ratio of 3:1 and the depth, the vessel at lot N 5 was perhaps suited for coastal trade on the Waddenzee and the other one with a length/beam ratio of 6:1 and a depth of only 0.8 m for the inland waterways.

A third example excavated as early as 1945 on lot M 107 in the Northeastern polder shows the same features as the other ships. Only the bilge, at the transition from the bottom of the sides is less angular (Fig. 3.6).

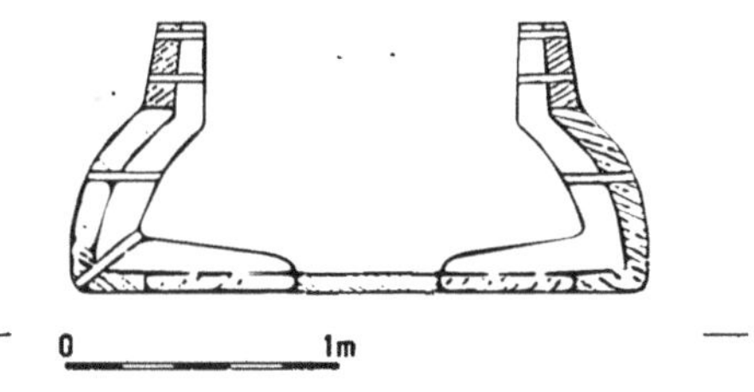

Fig. 3.3 Meinerswijk 3

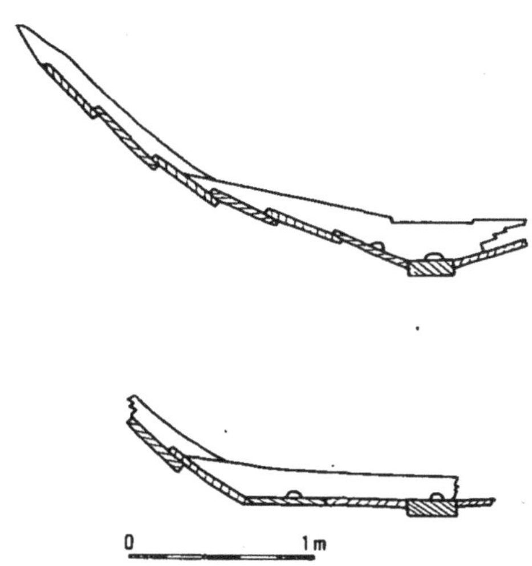

Fig. 3.4 Flevoland N 5

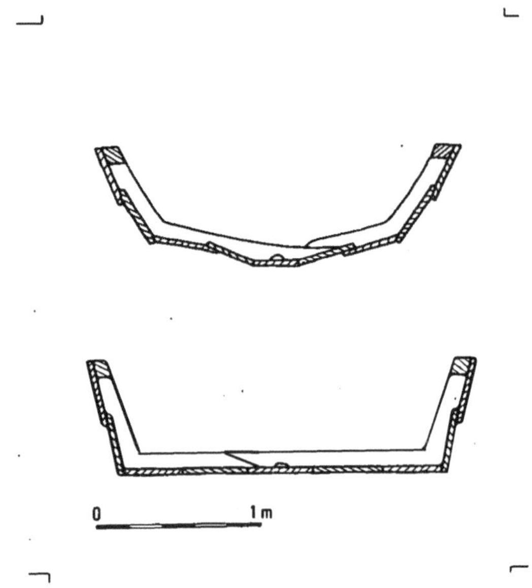

Fig. 3.5 Flevoland K 73/74

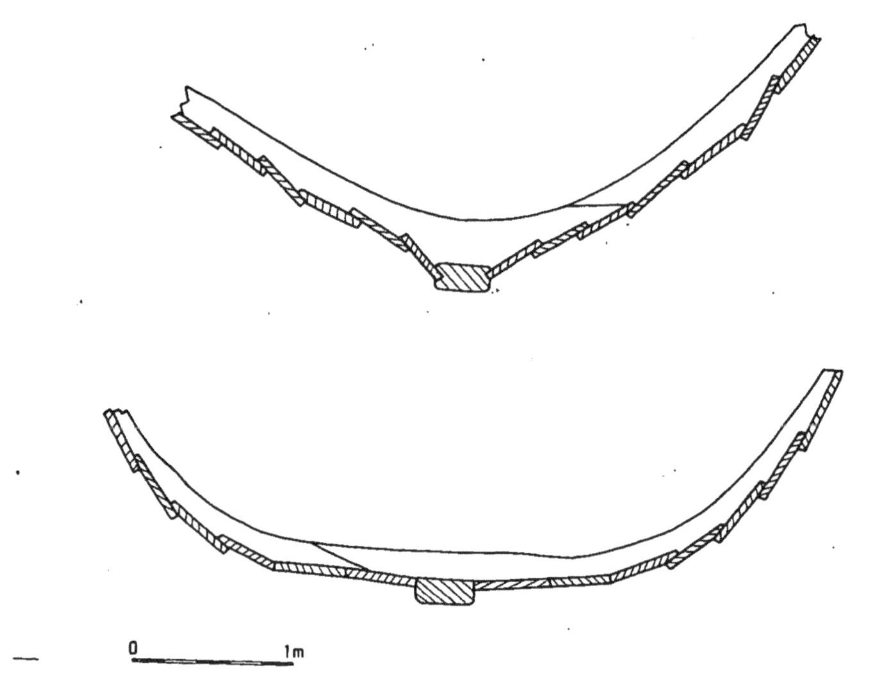

Fig. 3.6 Northeasternpolder M 107

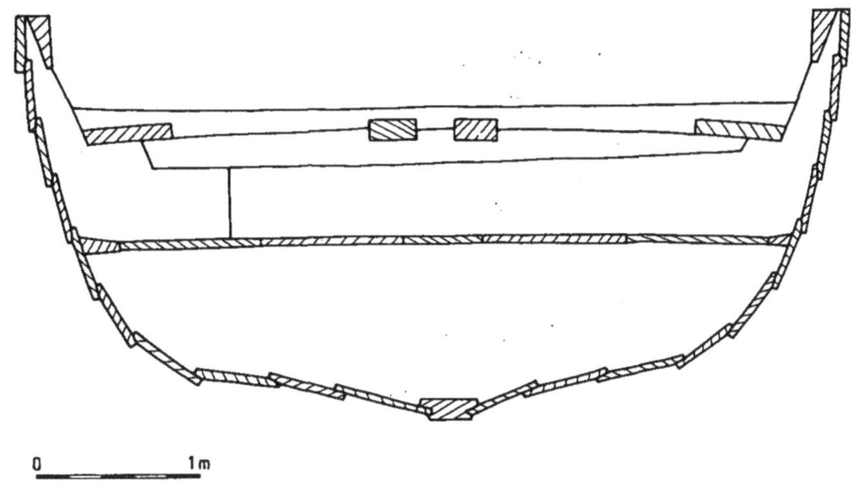

Fig. 3.7 Flevoland MZ 22

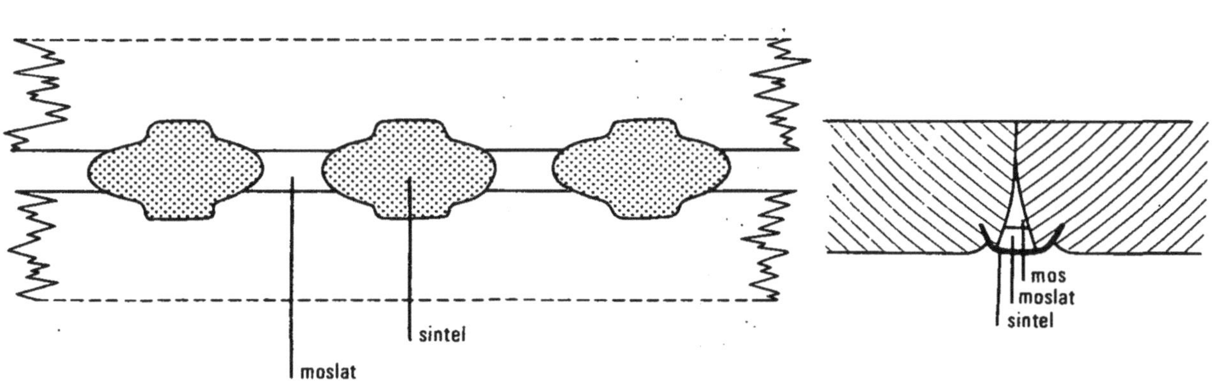

Fig. 3.8 Caulking method—breeuwnaad met gesinteld werk

Considering the dimensions and the date of wreckage (table 1) it is perhaps better to speak of a shipbuilding tradition of flat-bottomed vessels with several types for different purposes, rather than to speak of cog finds in the Netherlands.

Table 1 Dimensions of some mediaeval ships (in m)

Ship	Length		Width		Depth	L/B ratio	Date of wreckage
	bottom	overall	bottom	abeam			
Q 75	8,65	-	-	-	-	-	1325
N 5	12,50	14,50	2,00	4,50	1,00	3:1	1350
M 107	12,50	15,50	-	4,50	1,40	3:1	1400
K 73/74	-	15,80	-	2,40	0,80	6:1	1475
Bremen	-	23,50	-	7,50	5,30	3:1	1380

THE WATERSCHIP

Lastly I ask your attention for a large fishing vessel, called: waterschip. Data about this type of vessel range from the Middle Ages— about 1350—until the first half of the 19th century. These vessels were originally used to convey live fish; in later times for fishing and towing. The waterschip is particularly known as a vessel used for towing sea bound ships. In order to reach the Noordzee from Amsterdam these ships had to pass Pampus, shallows near Amsterdam. After the digging of the Noord-Hollands kanaal in 1829 the route across the Zuiderzee was no longer required and the waterschip became obsolete.

About six waterschepen have been excavated in the IJsselmeerpolders, dating from the 16th and 17th centuries. These ships have been built on a keel and show a S-shaped profile. They all had a bun, a compartment amidships open to sea water by means of small holes in the hull, to keep the fish fresh (Unger, 1978:31). Considering the shape, the strong construction and the division there is no essential difference between these vessels from the 16th century and the ship models of the 19th century waterschip.

In 1978 we had the opportunity to examine a late mediaeval waterschip, probably from the 15th century, found at lot MZ 22. Again there was no difference in shape and construction, with one exception: the hull was clinker built instead of carvel built. Figure 3.7 shows a cross section of this vessel at the front of the bun.

CAULKING METHOD

There is one element that these mediaeval ships have in common: the caulking method. All seams at the joints of both the carvel-built and the clinker-built planks show the so-called gesinteld werk (Fig. 3.8). The joint of the planks is widened and this breeuwnaad is filled with moss, overlapped by a lath and held in place by a row of iron clamps or wooden pegs. The

Dutch shipbuilders used the words: mos (moss), moslat (lath), sintel (iron clamp) and prik (peg). The mediaeval use of the word sintel is ascertained by a written source of 1402 (Jappe Alberts, 1967:164).

Considering the mediaeval ship finds in the Netherlands we can say:

- the archaeological evidence corresponds and is complementary to the finds known in Germany
- there is a large category of flat-bottomed vessels with several types for different purposes
- Q 75 dates from the 14th century
- the shape and construction of the waterschepen show that Dutch shipbuilding was not restricted to flat-bottomed vessels
- the caulking method with sintels was common in mediaeval times, but disappeared in the 16th century.

REFERENCES

Ellmers, D., 1972, Mittelalterliche Handelsschiffahrt in Mittel- und Nordeuropa. Neumünster.

Hoekstra, T., 1975, Note on ancient ships found at Utrecht. International Journal of Nautical Archaeology, 4:390-391.

Jappe Alberts, W., 1967, De tolrekeningen van Lobith over de jaren 1404/1405 en 1408/1409'. Bijdragen en mededelingen van het Historisch Genootschap, 81:58-177.

Sopers, P.J.V.M., 1974, Schepen die verdwijnen. Amsterdam.

Unger, R.W., 1978, Dutch shipbuilding before 1800. Assen.

van der Heide, G.D., 1974, Scheepsarcheologie. Naarden.

KEYWORDS

Medieval.

Netherlands; Zuiderzee; R. Rhine.

Boats with logboat elements; flat-bottomed vessels; waterschip.

Boatbuilding characteristics; caulking methods; dating.

4. A PRESUMPTIVE GERMANIC HERITAGE FOR A PORTUGUESE BOAT-BUILDING TRADITION

Octávio Lixa Filgueiras

Summary

Geographical features dominated Portugal's settlement pattern, and this accounts for the diverse cultural areas. The River Douro is a frontier between Mediterranean-style boatbuilding to the south and Nordic to the north where there are also differences between coast and river.

The cultural influences which acted on the northern region are investigated; in particularly, the Germanic invasions of the 5th century A.D., the Norman invasion of the 11th century, and subsequent medieval migrations are considered.

Situated on the SW end of Europe, Portugal is a contrastful country. With its 88.619 km^2 (about 1/5 of the whole area of the Peninsula) it is nevertheless an example of the regional diversification which characterizes both Spain and Portugal:[1] these differences are due not only to geographical features but also to ethnic-cultural roots. In spite of its remoteness and of the Pyrenean barrier the Peninsula bears the traces of all the great terrestrial migratory movements, and its situation between the Mediterranean and the North Atlantic has acted as a strong counterpoint between the terrestrial and the marine components—this last one being of great importance in the settlement starting from the coast.[2]

This marine component is in Portugal of the utmost importance due essentially to three basic circumstances: the disproportionate extent of the coastline (485 km) in comparison to the dimensions of the country (561 km long and between 218 and 112 km wide); its situation at the SW corner of Europe;[3] the ease of entrance granted by its principal rivers, with drainage-area amounting to 298.521 km^2 whereas Spain has only 284.979 km^2 with a land area of 492.320 km^2 [4] (Fig. 4.1). This is the reason why communication in the hinterland, up to the beginning of the 20th century, was by rivers ending in harbour-towns, plus connections between these towns by coastal navigation.[5]

In the history of settlement intensively dynamic periods naturally alternated with slowing-down or even stagnation phases: the situation of extreme <u>finisterra</u> may be considered as a landing beach (or launching site) as well as a hazy quay of oblivion....[6]

In Portugal, of all the regional differences, the strongest is the one which contrasts the Atlantic with the Mediterranean.[7] The R. Douro appears as a

total areas

Portugal 88.619 Km2

Spain 492.320 Km2

rivers drainage areas

Portugal 298.521 Km2

Spain ·284.979 Km2

dimensions of the country

561 Km long.

218/112 Km wide

845 Km ext. of seashore

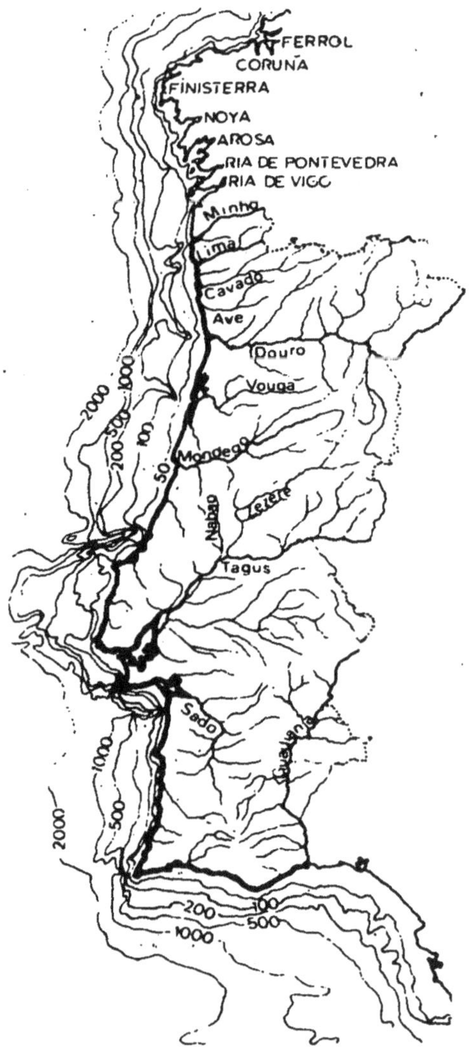

Fig. 4.1 The rivers of Portugal and Galiza

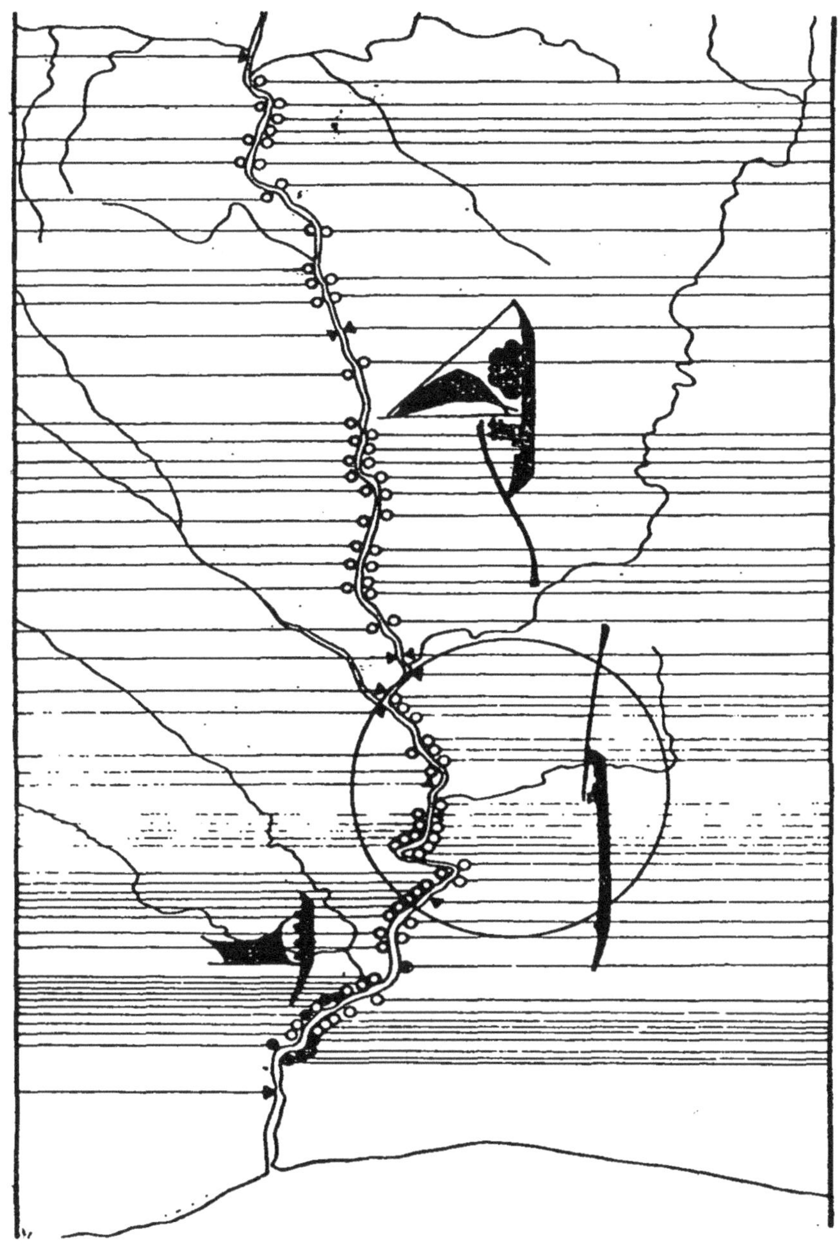

Fig. 4.2 River Douro; operating areas of sub-types belonging to the same family of traditional boats

natural frontier, being the second largest river in the Peninsula, and it is an absolute frontier when traditional boats are concerned.[8] South of this river in the central area of the country planked-canoes of Mesopotamian type[9] are dominant, whilst further south other models which originated in the Mediterranean of more or less recent times are to be found (see Figs. 9.1 and 9.7 in Filgueiras, 1977).

In the R. Douro and further north the representative boats reflect North-European cultural influences, the essential mark of the North West. Such influences are widespread in time and in origin.[10] This explains the great differences between the boats of the NW rivers (Minho, Lima, Douro) and the "lancha", the type adopted by all the coastal sea-fishing settlements in the area. (See Figs. 9.1 and 9.5 in Filgueiras, 1977.)

Very similar to the Breton "lanche", and in spite of being already integrated in the skeleton technique-carvel building system,[11] the NW lancha belongs to a specific context—the tradition starting with the settlement of the coast region and the reorganization of fisheries after the Norman and Moorish raids.[12]

The NW river-boats, the only ones in the country to illustrate the shell technique-clinker building system, raise peculiar problems of archaeological interpretation,[10] and these are considered below.

None of the boats of our NW rivers manage to escape the rule of contrasts. Although they are all linked by common features of construction, their types differ according to their origin: the *carocho* on the R. Minho; the *barco* or the *barquinha* on the R. Lima; on the R. Douro a large family of boats is to be found divided into two branches—the *rabelos* and the *valboeiros*.[10 & 13] As we shall see later on, there are not only formal aspects which individualize them but also something which touches their very structural conception. All those facts are obviously of the greatest importance and interest to the archaeological analysis.

Boats of the R. Douro

The family of Douro boats is the one which offers a best documentation of a building-pattern. The subdivision into two branches is due to the aggravation of traffic conditions beyond Entre-os-Rios (42 km from Oporto in the whole course of 200 km).[13-14] From Oporto to Entre-os-Rios, the navigation being less difficult, the hulls show stiffer lines, hard, almost aggressive forms (Fig. 4.2). Further on and up to Barca de Alva the thousands of dangers presented by an untamed mountain river caused the bow and the sides to be curved and the draught to be drastically reduced. This is well demonstrated when 18th century pictures and modern boats are compared.

In spite of the dangerous and difficult navigation, the R. Douro was for centuries the principal route for the transport of goods and people to and from the mountain hinterland. It did not suffer much with the advent of the railway and it was only in the 1950s that the truck carrying trade succeeded in dethroning it. It is, therefore, not surprising to find so many different models of boats (Table 4.1).[10 & 15]

Fig. 4.3 Several typesof R. Douro boat: barco rabelo (1st from the right); barquinhas rabelas (2nd and 4th from the right); rabão de apègadas (3rd from the right); Porto Arntigo (author's photograph)

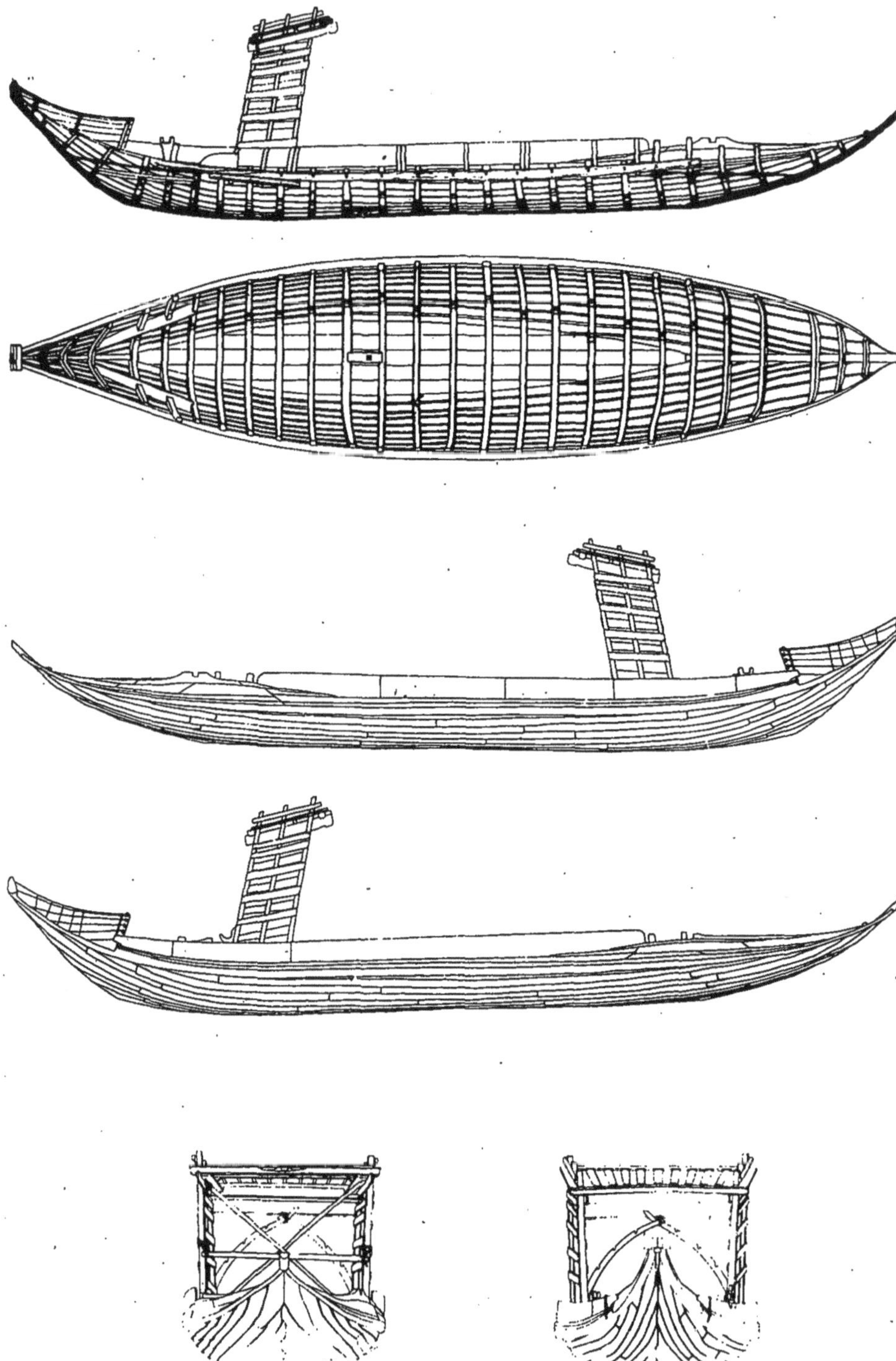

Fig. 4.4 Barco rabelo (plans by J. Cruz Gomes)

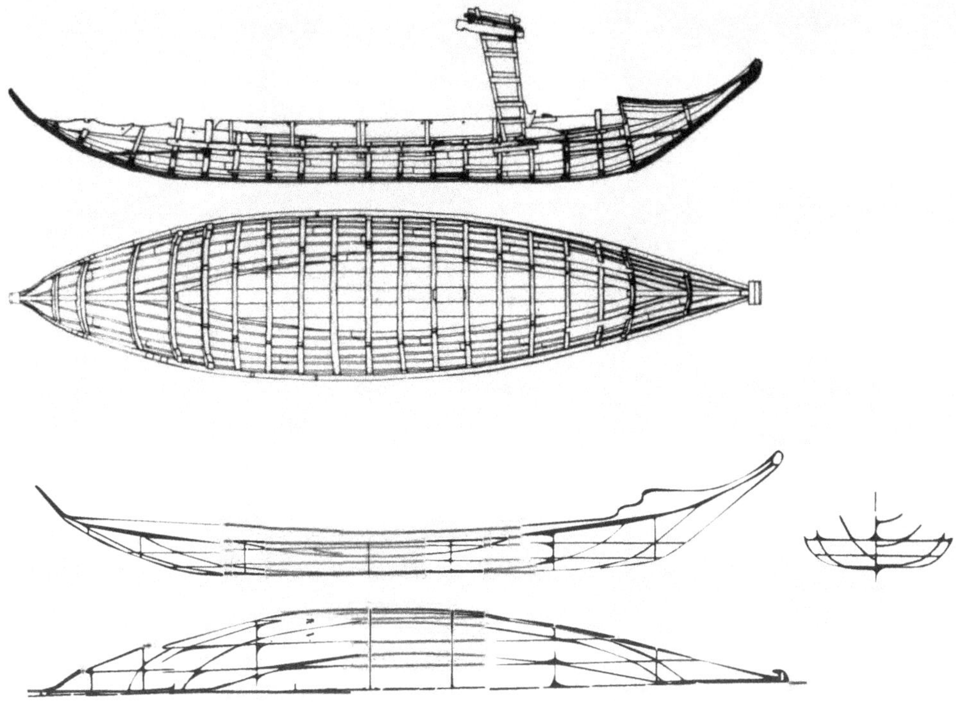

Fig. 4.5 Rabão de apègadas (plans by J. Cruz Gomes)

Fig. 4.6 Rabão de rabo baixo (top) and rabãozinho (below). Bitetos (photograph author)

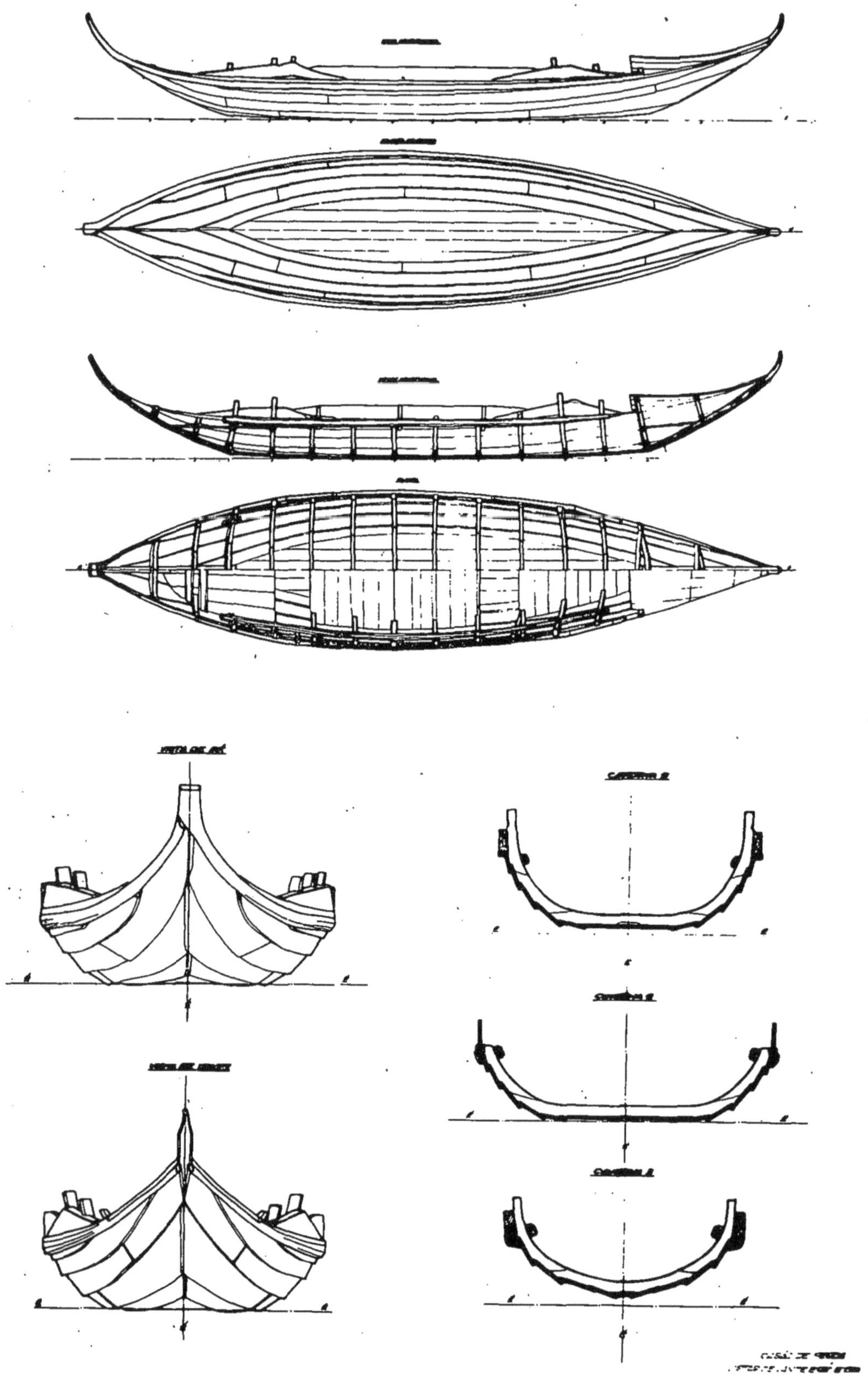

Fig. 4.7 Rabãozinho (plans J. Cruz Gomes)

Fig. 4.8 Rabão Carvoeiro; on the left two rabãoezinhos and two raboes de apègadas. Castel de Paiva (photograph author)

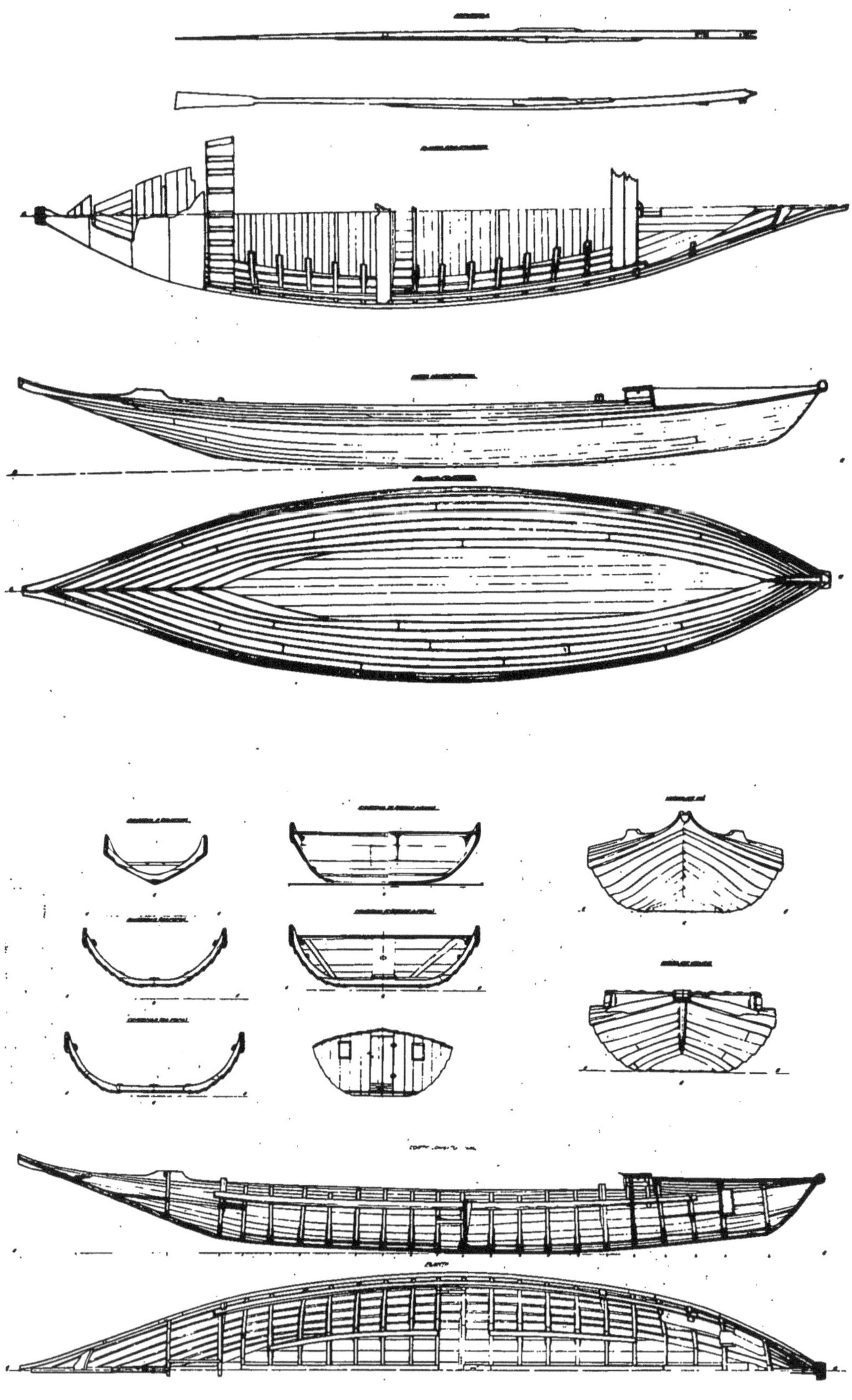

Fig. 4.9 and 10 Rabão carvoeiro (plans J. Cruz Gomes)

Fig. 4.11 *Barco de frete.* Casa Branca (photograph author)

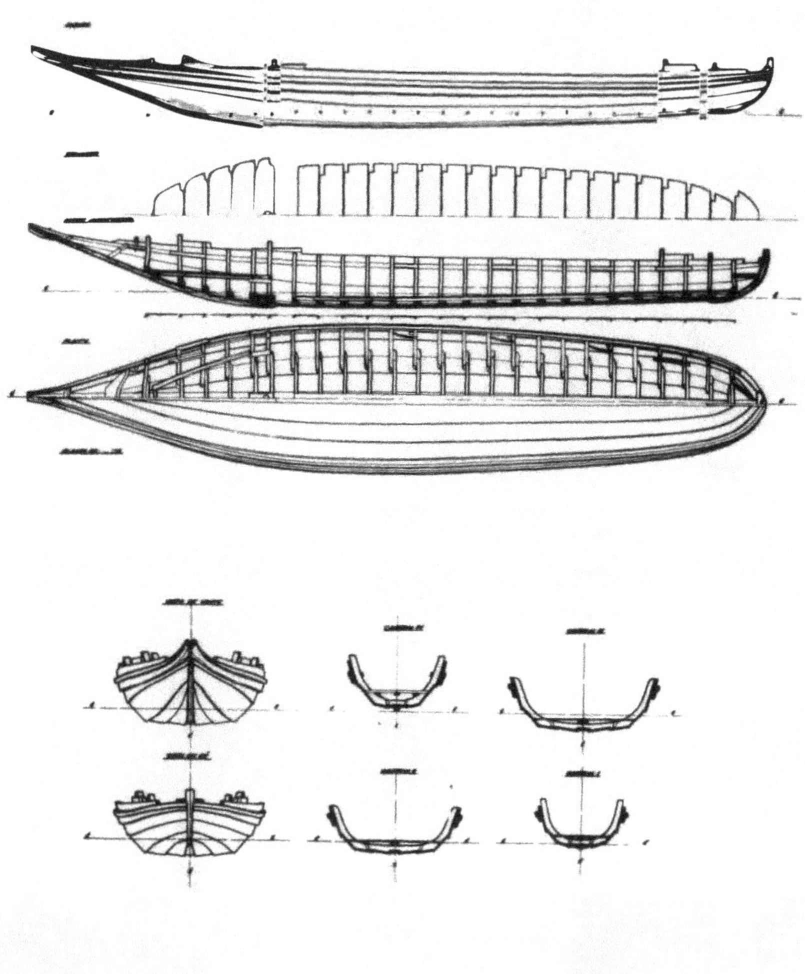

Fig. 4.12 *Barco de Areia.* Cantareira (plans J. Cruz Gomes)

Fig. 4.13 Valboeiro. Castelo de Paiva (photograph author)

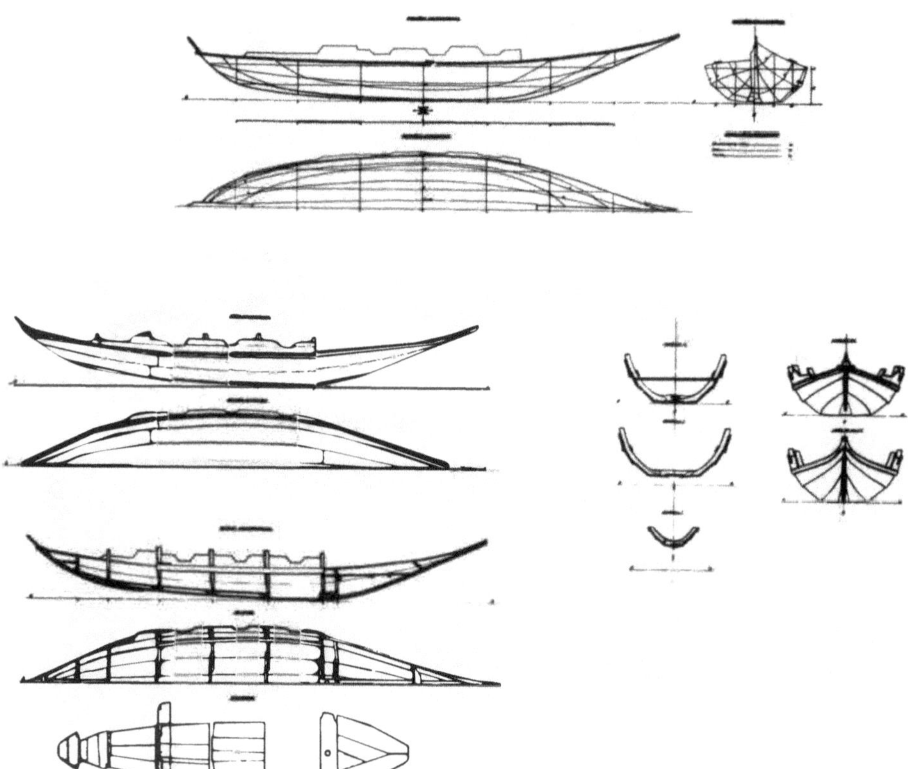

Fig. 4.14 Valboeiro (plans J. Cruz Gomes)

Fig. 4.15

4.16

Fig. 4.17

Fig. 4.18

Fig. 4.15 The rabelo building begins with the flat bottom of a lenticular plan. Castelo de Paiva (photograph J. Carvalho Dias)

Fig. 4.16 The shaping of the hull. Escamarão.

Fig. 4.17 Subsequently the ribs are fitted. Escamarão.

Fig. 4.18 The ribs are shaped one by one matched to the very places where they are to be placed. Escamarão. (Figs. 16, 17, 18 photographs author)

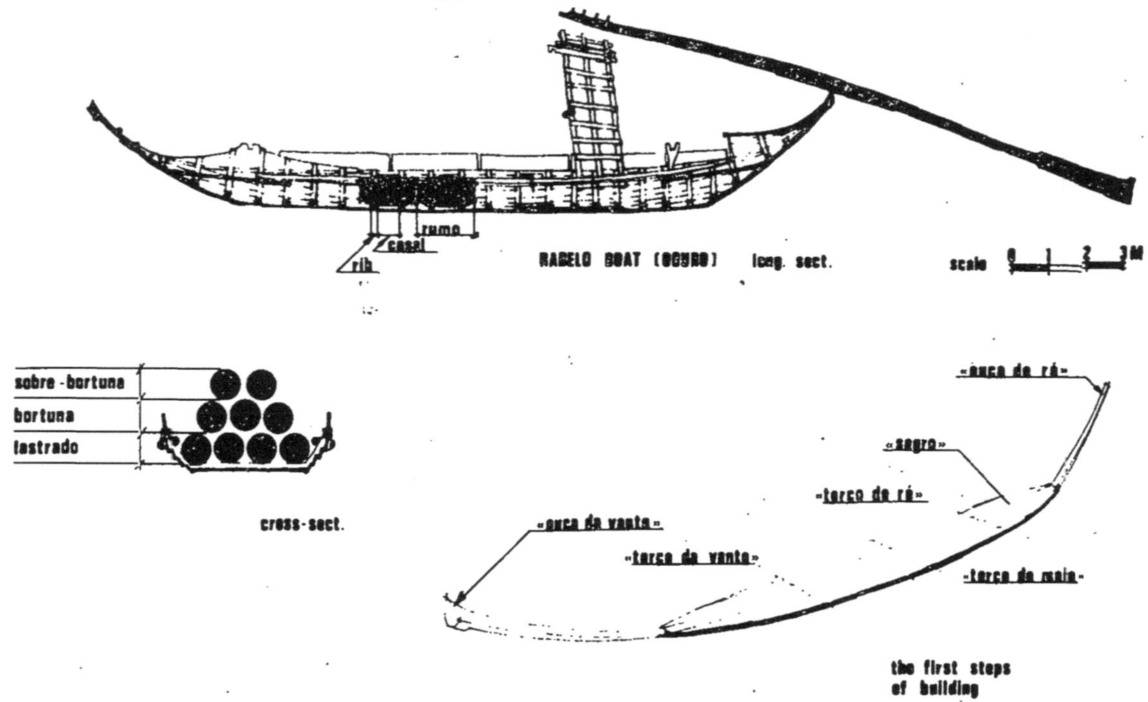

Fig. 4.19 Some of the "secrets" of rabelo building

Fig. 4.20 Lima Barca. Ponte do Lima (ancient post-card)

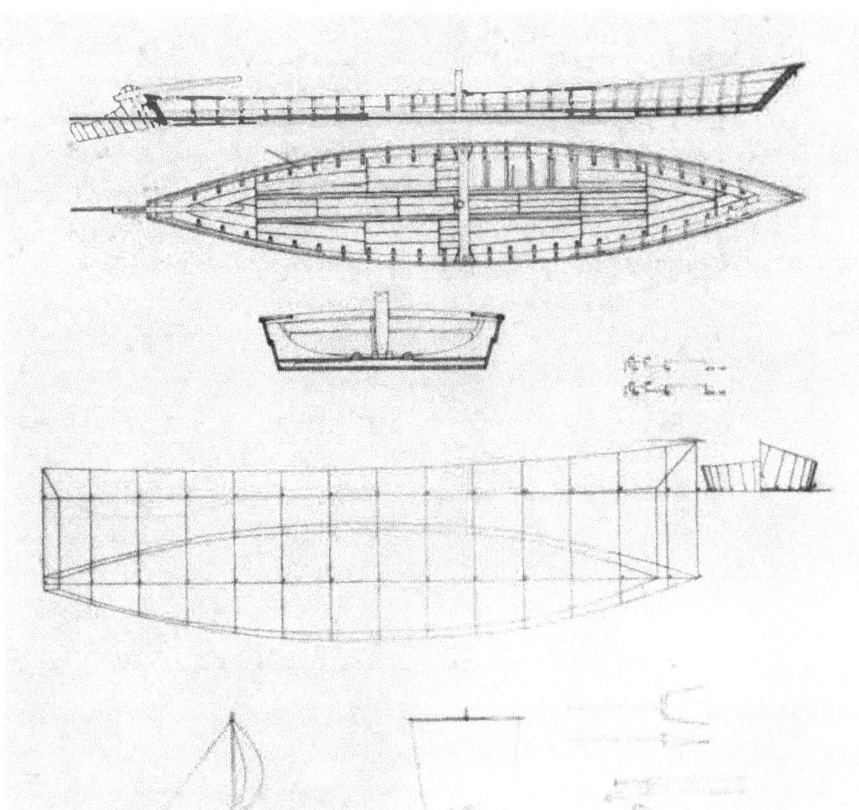

Fig. 4.21 Lima Barca. (Plans Jose Geraldes.)

Fig. 4.22 Lima Barquinha. Viana do Castelo (photograph author)

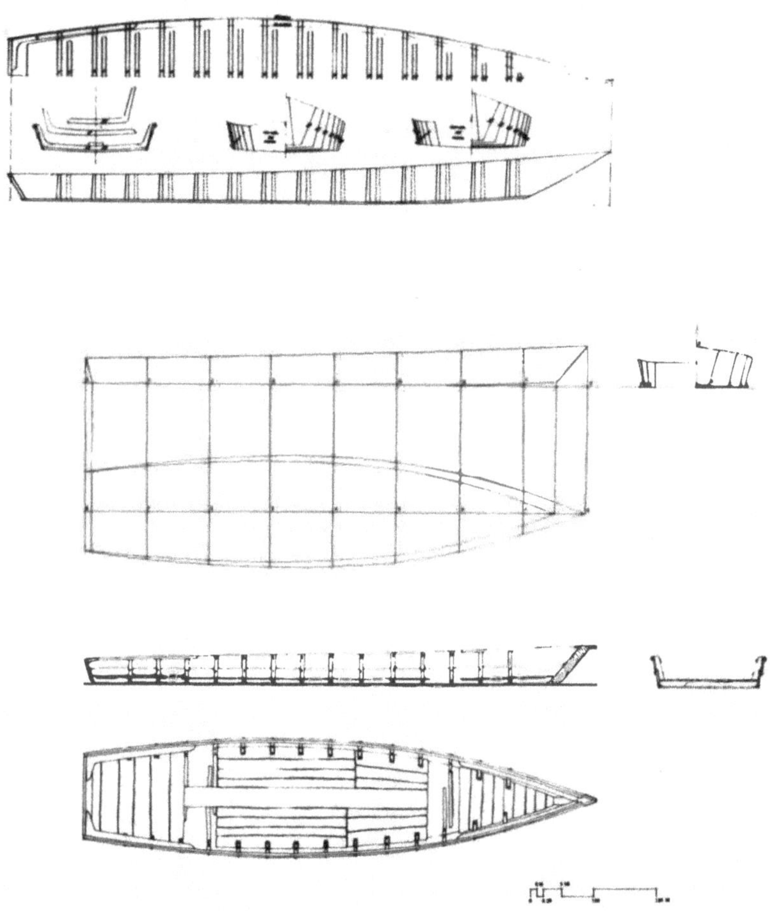

Fig. 4.23 Lima Barquinha (plans José Geraldes)

Fig. 4.24 Carocho. Seixas (photograph author)

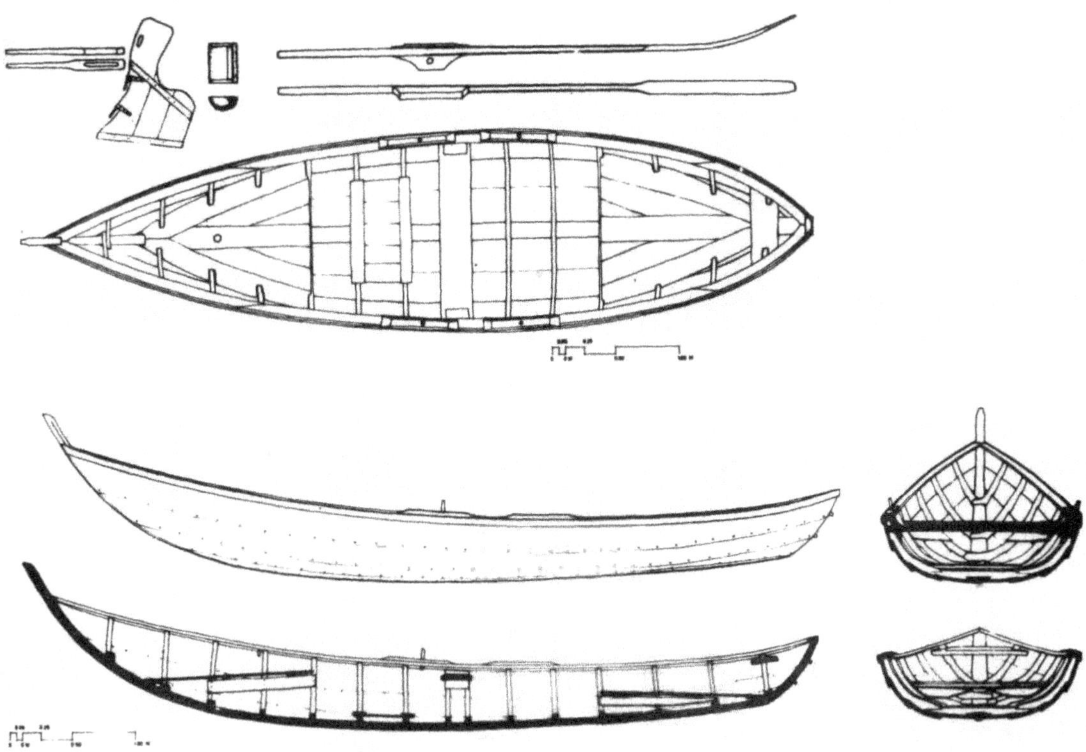

Fig. 4.25 Barco da Parada (plans José Geraldes)

Independently of a few details which distinguish them, all the boats of the R. Douro have a flat and keel-less bottom; bow and stern are built of blocks which are covered by the hull planking; the ribs are of the "long and short work alternating" type;[16] the planking is of the clinker type fastened by "nails hooked over so as to re-enter the planks"[17]—in perfect obedience to a pure shell/technique.

The building begins with the flat bottom of lenticular plan (Fig. 4.15) to which the inner pieces of reinforcement at bow and stern are fixed. Afterwards comes the shaping of the sides (Fig. 4.16)—and only at the end are the ribs fitted (Fig. 4.17), shaped one by one in the very places where they are to be secured (Fig. 4.18). Finally the whole is consolidated by guards and clamps.[15] To these details, important but by no means unique, may be added the "secret" of the building masters.

They do not use moulds: the conformation of boats is the result of a few key-measurements, many years of practice and the skill of craftsmen. The length of the bottom is a function of the required load carrying capacity (in casks). The bottom is divided into four parts marked by transverse lines, tercos,[18] which also give the width corresponding to the development of the specific lenticular form (Fig. 4.19). By steadying her on the bearers they make the bottom level (with water) as far as the aftermost terco; from here aft they bend the bottom upwards slightly.

Once the block stems are set and the line of the whole is dressed, her profile is corrected from far away by sight. When the hull is half built a first measurement verification is made. When the hull is completed the shape of the boat is verified by measurements taken from a line between the tips of the stem and the stern, and corrections are made when necessary.[15]

The interval between ribs is directly related to the way of loading the wine casks. Each cask is placed on two adjacent ribs, in the fore and aft direction, adjacent casks having their tops next to one another (Fig. 4.19). The stowage pattern conforms to given rules and therefore the positioning of the ribs has to be accurate so that the required number of casks may be carried (Fig. 4.19). We see that in the R. Douro region the ancient nautical building measure—the rumo—still exists: it is equal to double the distance between the two adjacent ribs. As there is a fixed relationship between the size of the casks and the length of the bottom, and the breadth and depth are fixed according to the type of boat, the rumo can be used to gauge the tonnage (tuns).[19]

The interest always raised by these boats, added to the knowledge of the details I have summarised makes it urgent to identify their genealogy. Unfortunately nothing is known about the time and the way they appeared. The first notice enabling us to identify them is quite recent—1712. It says "Familiar boat of the Douro having as rudder a large oar known as espadela and and two other oars on each side with which the boat is steered. Navigium quod remus pro clavo regit, vulgo Azurracha".[20] This word, Azurracha, no longer in use, can be found in an earlier document—1631—in which the reference to the boats is connected with the importance of river borne trade.[21] Its origin would probably be the medieval word azora, wine brand, in logical relationship with the principal use of these boats.[22]

But the importance of river navigation appears in older documents as, for example, in a permit of 1252 for the district of Penajóia to proceed to works in the river Douro provided they "non tollant passaginem Barcas que par riuum passauerint";[23] the argument between the Crown and the Bishops of Oporto (documents of 1254 and 1284) when both of them wanted to " receive the maximum of the taxes applied to river borne trade (...) which was really important and intense";[24] the proved existence of barcas taverneiras intended to supply the crews of long distance boats, as still happened a few years ago.

In addition, for those who are acquainted with the riverside traditions, expressions such as "barcas de Riba Douro" (Upper Douro barges) reinforced by specification of employment are terms of typological reference: "mando vobis que de omnis Barcis et Barquetis qui venerint de ripa de Dorio cũ vinio et cũ lignis et cũ alijs rebus...".[26]

The rabelo is still nowadays called only barco (boat): barca is barca de passagem (ferry) of the largest sort; and barquinhas (little barges) the rabelos of the shortest dimension.[27] And when in a document of sharing of (river) harbour taxes between the Bishop and the Oporto Cathedral Chapter —A.D. 1200—"boats of four or more oars" are mentioned—"quatuor remorum et supra",[28] one is at once reminded that Alto Douro (Upper Douro) boats are rowed with four oars.

Considering the above mentioned aspects, the clear archaism of these boats and the conservatism of regional boatyards, the identification of medieval examples with types which lasted till our time is credible. There is something more: if we refer to the first known notice—A. D. 922, about river taxes—nabulum and portaticum—emphasising the antiquity of the importance of fishing and trade,[29] we are tempted to include in the same line of continuity the boats which permitted the exercise of those traditional activities. Nevertheless what should one think of the suggestion that the tradition of a boat building industry in those times originated in the Roman ages?[30] The evidence is the granting of a passage between Portus and Cale on the road Lisbon-Braga (Braga being an important focal point in the Roman military road net): but this is not sufficient. We have to appeal to the "testimony" of the boats themselves. However, except for the fact that both Greeks and Romans employed the shell/technique[31] nothing allows us to connect their boats to that of the River Douro boats.

We are in a region once occupied by the Celts from the 5th century B.C.;[32,33] however other evidence indicates that their direct influence on boatbuilding was minimal. Strabo states that the natives used to navigate in the kallaikói area[34] in skin-boats and were beginning to employ dugouts.[35] When the same author declares that the river is navigable by large ships—magnis scaphis—he has surely based this on other peoples' experiences, namely upon the ones who were interested in trade, the Phoenicians for instance.[30] And by which ships? The R. Douro boats present an unquestionable Nordic appearance as well as the use of shell technique-clinker building. However, there are other features: stem and stern replaced by inner reinforcement pieces (blocks); the type of ribs; the characteristic longitudinal profile; the type of the rowlocks, and so on (Table 4.4).

Comparing these technical details with the general evolution of local settlement, we come to two hypotheses for the origin of these boats: the

Suevi,[36] or the Visigoths.[37] The idea of Viking influence[38] is to be put aside because of the archaism of our boat types.[39] The Germanic invaders who settled for centuries in this R. Douro area were the Suevi (5-7th centuries) followed by the Visigoths (see Fig. 9.4 in Filgueiras, 1977), until the Arabic domination (in 997 Almansor conquered Oporto arriving from the south with troops coming by land and by sea).[40] While the Visigoths had an important effect on the social-political organization (Law, Rights, Religion), the Suevi reached more directly and more deeply the cultural roots of a population which had a strong feeling against Romanization. After "the archaic roughness" of the Iron Age culture (5th century B.C.-5th century A.D.)[41] the most natural transformation came through these people:[42] an example is the adoption of the plough of Suevic type.[43]

A process of analytical-systematic comparison[44] could clarify the origin of these boats but we have insufficient material at present.[45]

Boats of the R. Lima

Passing to another river, the Lima, about 70 km further north, we find ourselves in quite a different landscape. It is no more a violent mountain stream but a quiet and poetic river—the river of oblivion—the Lethes, as the Celts may have christened it.[46] This river has been an important trade route, especially in the Middle Ages and up to the middle of the 15th century.[52] Its boats, whose lines are much more primitive, are also flat-bottomed and shell technique—clinker built. They remind us of the Somerset turf boat[47] except for a rabbeted stem like the ones in the East Germanic Baltic coast boats;[48] a transom aft; and ribs—"long work, setting by pairs", one for each side—evidently similar to the Rhine boats studied by Ellmers.[49] Another curious detail is the wooden truss, resembling the ones in the Pomeranian boats[50] (Table 4.2).

If it were not for Strabo's reference (above) we would be tempted to link these boats with the most primitive Celtic types, a boatbuilding tradition which lasted in Europe well into the Middle Ages.[51]

Boats of the R. Minho

If the distance separating the Lima from the Minho is shorter—24 km—greater is the difference between the boats of this last river—the carochos family (Table 4.3)—and the ones we have been dealing with.

In the carocho the clinker built system is mixed with the skeleton technique. This hybridism is attributable to the influence of the maritime coast shipyards. Structurally and formally she presents clear improvements in comparison to the other specima. Being a double-ended model, she has the aspect of an insect's shell which is the origin of her peculiar name (carocho being the Portuguese name for a kind of a bug). It is as if something of a curragh[53] had been transposed to a Nordic wooden body—earlier phase of the Oselver boat?[54] She is a keel-boat, but with a flat-bottom giving a light draught. And while each rib is formed of floor timber plus futtock, her mid-ship section presents an unquestionable interest mainly for her T-shaped keel[55] (Table 4.4 & Fig. 4.26).

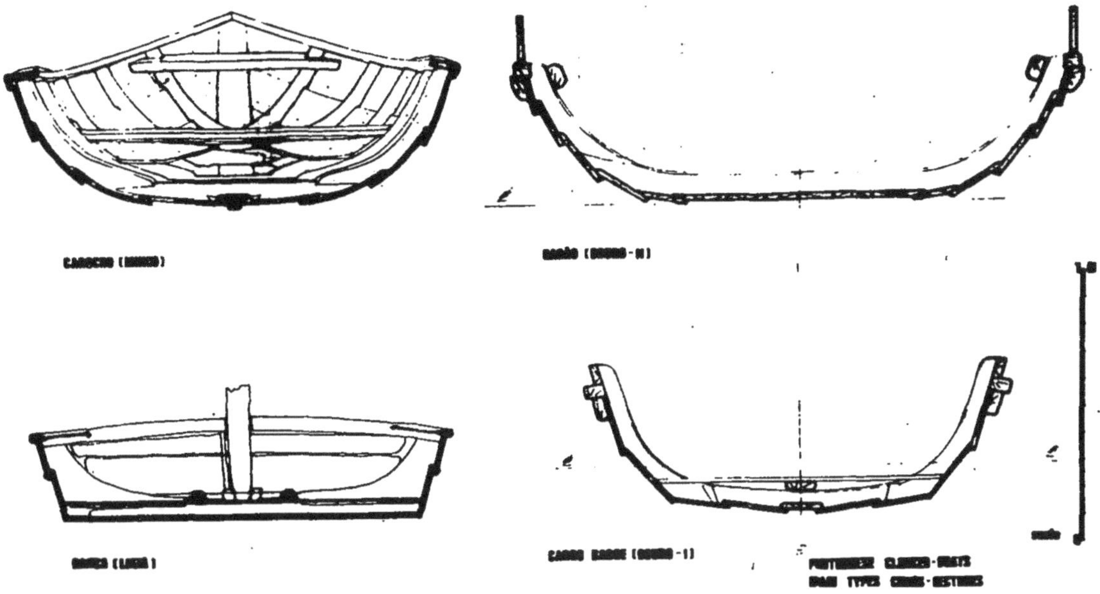

Fig. 4.26 Transverse sections of Minho (carocho), Lima (barca) and Douro (rabãozinho and barca de areia) boats

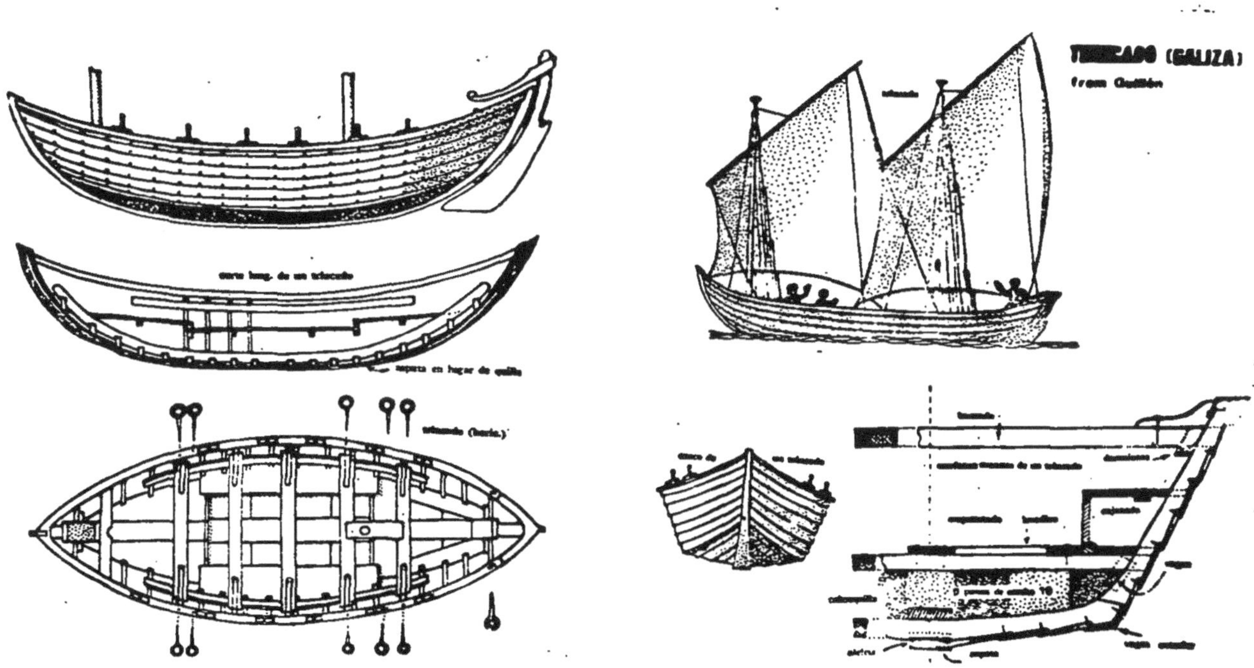

Fig. 4.27 Galiza, Trinicado (plans Guillén)

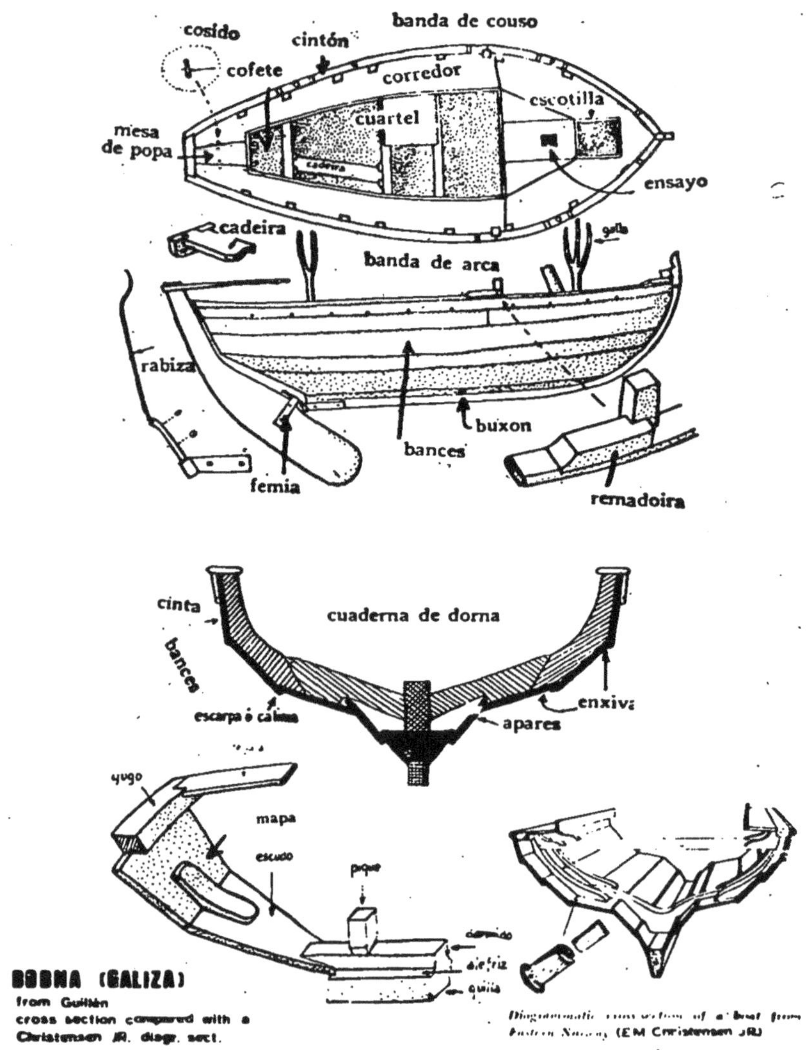

Fig. 4.28 Galiza, Dorna (plans Guillén) and diagrammatic section of a boat from Eastern Norway (after A. E. Christensen)

Such an unusual type in comparison to all the others in the NW rivers makes us extend our analysis to other similar models in Galiza, Spain. Julio Guillén in his essay bearing the meaningful title "The large family of Trincados and other regional boats presenting Norman influences" refers to the barcolongo, trincadura, trincado, traina, dorna.[56] All of them have disappeared except for the dorna (Fig. 4.28), still employed on the rias of Arosa and Pontevedra; el Grove and Sta. Maria de Lanzada. Guillén might also have mentioned the trincado[57] (Fig. 4.27) from the rias of El Ferrol and la Coruna for there was a model in the Madrid Naval Museum. Both of them seem to be more developed than the carocho, especially the dorna in her general lines, midship section[58] and type of keel. These boats are built by the shell-technique system. The trincado herself in spite of a few details such as having a bottom plank instead of a keel, and the L-strakes,[59] also show improvements when compared to the Minho boat.

To return to the problem of origins—or cultural influences— no document is known earlier than the one from the 15th century which deals with the dorna. However, Farina Busto's interesting essay advances the hypothesis that the dorna is from the 13th century.[60] Related to the dorna and the trincado are other developed boats which went out to sea, including the Caminha barco da ribeira, a developed model of the carocho engaged in coastal fishing. A text of A. D. 1112 reminds us of a meaningful situation—"fluvium quod Mineum vocatur per quod Sarraceni e Normandi frequentius solebant intrare".[61]

Would the appearance of this sub-group have something to do with the Norman influence after the phase of violent contacts? The Normans appeared for the first time during the kingdom of Ramiro I (842-850), their most famous attacks being during the kingdom of Ordonho I (850-866), Afonso II (866-910), Afonso V (925-931) and Ramiro III (966-984). But in the 11th century they were apparently already engaged in "friendly" relationship with the local population, as shown by the help given to Count Veremundo Vergilaz in the fight against Basque rebels entrenched in the Castle of Lapio (Lugo—Bermudo III's age 1028-1037): "Coadunavit scipse Comes, cum omnes suos Barones & cum Gens Leodomanorum et cerravit ipsa pena & pressit ea per fortia, & cremavit & solavit ea".[62]

CONCLUSION

For the time being we can go no further. In short, the diversification of the NW Peninsular boats of presumed Nordic descent makes us suppose there to have been several influences and different culture-diffusing agents. As we have seen, Strabo's reference to the leather-boats and dug-outs used by the natives in this area seems to contradict the hypothesis of a Celtic cultural component.[63] On the other hand, the depredatory feature of the Vandal's maritime raids does not favour another line of research.[64]

The permanency of the Suevi for two centuries (5th-7th B.C.) in the NW and the kind of relations established with the local population support the thesis that they could have introduced (even having come by land) techniques such as the building of river boats (Douro?, Lima?)—the Suevi rather than the Visigoths. However, we lack evidence to prove this influence as far as boats are concerned. The evolution shown by Galiza boats—from the Minho to El Ferrol—introduces the possibility of intervention by other Northern peoples—namely the Normans.

But being in an area subject to multiple "encounters", 65 the medieval melting-pot could have resulted in other contributions the search for which has lost us for the moment in a mist deeper than the one enveloping the matter of Britain:

> "Ne sont que iii matières a nul homme antandant;
> De France et de Bretagne et de Rome la grant;
> Et de ces iij matières n'i a nul samblant..."
>
> Jean Bodel Chanson des Saxons
> (13th century)

Acknowledgement

The English version of this paper was prepared by Luisa Feijó.

NOTES

1. O. Ribeiro, 1955, pp. 9-12.
2. O. Ribeiro, 1977, pp. 47-
3. O. Ribeiro, III, pp. 434-435; J. Cortesao, pp. 30-31; A. Schulten I p. 308 and ff.
4. F. Castelo-Branco, pp. 60-61.
5. O. Ribeiro, 1977, p. 63; J. Serrao, pp. 192-194; W. M. Kinsey, pp. 312-313.
6. O. Ribeiro, 1955, p. 67; III p. 435.
7. O. Ribeiro, 1977, p. 91.
8. Under other aspects, "rivers are not absolute boundaries"; O. Ribeiro, 1977, p. 76.
9. O. Filgueiras, 1977.
10. O. Filgueiras, 1977b.
11. O. Filgueiras, 1966.
12. O. Filgueiras, 1970a.
13. O. Filgueiras, 1959, p. 7.
14. O. Filgueiras, 1962.
15. O. Filgueiras, 1956.
16. E. McKee, Fig. 4 "O".
17. E. McKee, Fig. 2 "E".

18. Literally: the <u>thirds.</u> This is not a referance to the division of the bottom into three equal parts—but to the three transversal lines which divide the bottom into four equal parts.

19. O. Filgueiras, 1961.

20. Bluteau.

21. Novaes, II pp. 241/242.

22. A. Nascimento pp. 276/277.

23. F. Castelo Branco, p. 43.

24. F. Castello Branco, pp. 43, 45.

25. M. A. C. Pico, p. 50 ref. docs. from 1255 and 13 ; cf. also Viterbo.

26. O. Filgueiras, 1959, pp. 12, 12.

27. 1254 cit. M. A. C. Pico, p. 34; F. Castelo Branco, p. 44.

28. A. Sampaio, 1923, a, p. 339.

29. A. Sampaio, 1923 b, pp. 378-383 (in spite of P. David's reticence about the doc. ref. S. Soares IV, p. 454).

30. A. Sampaio, 1923b, p. 383. Long before the Romans "the Phoenicians were engaged in river borne trade with the agricultural peoples of the Meseta(...) thus showing the existence of a double-sense traffic of which <u>Cale</u> was the center and stimulum". O. Ribeiro, 1977, pp. 100-101.

31. O. Hasslöf, p. 49 & ff.; A. E. Christensen Jr., 1977, p. 276.

32. M. Cocheril, p. 57 & ff.

33. A. E. Christensen Jr., 1977; P. Marsden, 1976 and 1977.

34. Spanish edition by A. Garcia Y Bellido, p. 136.

35. O. Filgueiras, 1962; cf. F. Alonso Romero; also note 51.

36. O. Filgueiras, 1964, pp. 33, 40-44.

37. F. Beaudouin, 1964, pp. 80-86.

38. The Normans appear for the first time in the Peninsula during the kingdom of Ramiro I and their raids go on until the 11th century: A. Sampaio, 1923a, pp. 286-289, A. H. Oliveira Marques, pp. 30 & ff.

39. The remark that the Suevi were not sailors (A. Sampaio, 1923b, pp. 270-271) is not relevant for what is being discussed is not Atlantic navigation but river navigation and, above all, a technological contribution (boat Building).

40. A. Sampaio, 1923b, pp. 378, 379.

41. O. Robeiro, 1955, p. 71.

42. O. Ribeiro, 1955, p. 76; III p. 441. Even if W. Mathes' interpretation would be nowadays taken into consideration (I, pp. 49 & ff.) it deals with former periods (Bronze Agé—early Iron Age—cf. Brøgger & Shetelig,

pp. 34 & ff.). For the present case the possibility of a relation between pre-Viking boats, still keel-less, and the ones of the lower part of the River Douro whose flat bottom seems to be reduced to a plank, has to be verified.

43. J. Dias.

44. E. McKee, pp. 3-12.

45. A. E. Christensen Jr., 1977, pp. 276, 277.

46. M. Cocheril, p. 55, ref. Pliny (Natural History, L. 4, 22 ch.) appeals to Strabo's testimony about the Celtic origin of this tradition.

47. E. McKee, Fig. 2 "X" ; and Fig. 3 in B. Greenhill.

48. Doc. Kurt Kühn, Fig. 128 in B. Greenhill. In the Maritime Museum of Bremerhaven there is a boat presenting an identical bow—the binnenboot of Havelkahn; cf. D. Ellmers 1972b, p. 104, Fig. 74.

49. D. Ellmers, 1972a, pp. 47-48; and 1972b, p. 104, Fig. 74. In the Maritime Museum of Bremerhaven there is a Rhine boat with the same type of rib disposition—the dreibordrachen.

50. P. Smolarek, Fig. 100. In the same work, Fig. 31 presents the Falterbo prâm (Dr. R. Blomquist) whose basic lines are those of the barquinha— except for the detail of the L-planks. Unlike the Bevaix boat (Dr. Arnold) and the Zwammerdam boat (Dr. P. Weerd & Haalebos) which derive from dugouts (P. Marsden, 1976; cf. B. Greenhill, ch. 8), the Lima boats are of the planking-canoe type as defined by D. & G. Ibarra-Grasso, pp. 167ff. The comparison above may be extended to boats of other areas, like the ones of the R. Adour, cf. F. Beaudouin, 1970, esp. pp. 111ff. Cf. A. E. Christensen Jr., 1977, p. 276.

51. P. Marsden, 1977, p. 282.

52. A. Sampaio, 1923a, pp. 314-318 and 354-355; F. Castelo-Branco, pp. 41, 55, 61.

53. J. Hornell, p. 146.

54. B. Greenhill, Fig. 156. Cf. the Nordmøre boat in A. E. Christensen Jr., 1968, p. 81.

55. "O" and "F" intermediate model, Fig. 1 in E. McKee.

56. J. Guillén, pp. 13-24.

57. J. Guillén, pp. 8-12.

58. The comparison is striking with a boat from Eastern Norway in A. E. Christensen Jr., 1968, p. 78.

59. See second part of note 50. But we cannot forget that available documentation is only a model: how far is it a safe basis for study especially for the midship section? The fact is that with a similar lengthways profile the fishing boat of the Faröer—the inseln boat—has a round cross section and an acute keel.

60. F. Farina Busto, p. 302.

61. A. Sampaio, 1923a, p. 290; M. Mollat, p. 3.

62. A. Sampaio, 1923a, pp. 286-289. Cf. J. Mattoso p. 70 and note 51. On the difficult subject of the relations with the invaders, visitors and Northern emigrants, see also A. H. Oliveira Margues, pp. 29-51, 221.

63. Strabo's testimony does not correspond, of course, to visual evidence (cf. A. Garcia Y Bellido pp. 32-33). But we cannot forget the reservations made to the designation 'Celtic' applied to boats 'distinctively different from those that we tend to classify under the terms Scandinavian, Mediterranean' (P. Marsden, 1977, p. 283).

64. The Galiza coast was the object of frequent attacks. Idácio's notices refer the ones of 445, 456, 459 and 465 (note 22 in F. Faina Busto, p. 294, cit. EspanaSagrada). Cf. A. Sampaio, 1923a, pp. 270-271.

65. The Christian Reconquest and the formation of the Kingdom of Portugal are historical processes to which are linked immigration phenomena (from Europe to Galiza and Portugal) with the corresponding cultural incidences (cf. O. Ribeiro, III).

66. Quoted by James Douglas Bruce The evolution of Arthurian Romance from the Beginnings Down to the Year 1300 2nd ed. Gloucester, Mass. 1958 in A. C. Bruneti; note 1, p. 131.

REFERENCES

Alonso Romero, F., 1976. Relaciones Atlánticas Pre-Históricas entre Galícia y Las Islas Británicas y médios de Navegación, Ediciones Castrelos-Vigo.

Beaudoin, F., 1964. Les Bateaux du Douro, Études des Origines, Museu de Etnografia e História da Junta Distrital do Porto, Porto.

Beaudouin, F., 1970. Les Bateau de l'Adour, Genèse d'une Architecture Nautique. ed. Musée Basque, Bayonne.

Bluteau, R., 1712. Vocabulário, Coimbra.

Brøgger, A. W. and Shetelig, H., 1951. The Viking Ships, Dreyers Forlag, Oslo.

Bruneti, Almir de Campos, 1974. A Lenda do Graal no contexto heterodoxo do Pensamento Portugues, Soc. de Expansao Cultural, Lisboa.

Castelo-Branco, F., 1958. 'Do tráfego fluvial e da sua importancia na economia portuguesa' in Boletim da Sociedade de Geografia de Lisboa, Lisboa.

Castro, D. Jose de, 1943. Estudos Etnográficos—Aveiro II, Instituto de Alta Cultura, Lisboa.

Cocheril, Fr. Maur, 1970. 'Bretagne et Galice', in Pax, Chronique de L'Abbaye de Landévennée.

Cortesão, J., 1964. Os Factores Democráticos na Formacão de Portugal, Portugália Editora, Lisboa, (2nd ed.).

Christensen, A. E., 1968. Boats of the North, Det Norske Samlaget, Oslo.

Christensen, A. E., 1977. 'Ancient Boat Building—a provisional classification', in Sources and Techniques in Boat Archaeology, National Maritime Museum, Greenwich. B.A.R. -S29

David, P., undated. vd. Soares, Torquato de Sousa.

Dawson, C., 1960. Le Moyen Age et les origines de L'Europe (French translation of The Making of Europe), Arthaud, Paris.

Dias, J., 1948. 'Os Arados portugueses e as suas prováveis origens', in Revista da Universidade de Coimbra vol. XVII, Coimbra.

Ellmers, D.,1972A. 'Ein mittelalterliches Schiff aus dem Rhein,' in Die Heimat, Zeitschrift fur niederrheinische Kultur- und Heimatpflege, Krefeld 43.

Ellmers, D., 1972B. Frümittelelterliche Handelsschiffahrt in Mittle- und Nordeuropa, Karl Wach Woltz Verlag, Neumünster.

Estrabao. vd. = Garcia Y Bellido, A.

Fariña Busto, F., 1975. 'Sobre el origen histórico de la dorna', in Boletin Avriense, Orense.

Filgueiras, O. L., 1956. Rabões da Esquadra Negra, ed. 'O Pejão', Porto.

Filfueiras, O. L., 1959. 'Tipologia dos Barcos do Douro', in Boletim Cultural da Camara Municipal do Porto, vol. XXII, Porto.

Filgueiras, O. L., 1961. 'Rumo, palavra-chave da arte da construcão naval', in Revista de Guimarães, vol. LXXI, nos. 3-4, Guimarães.

Filgueiras, O. L., 1962. 'Da Navegacao do Douro', in Actas do I Colóquio Portuense de Arqueologia, Studium Generale, vol. IX, Porto.

Filgueiras, O. L., 1964. 'Entre Normandos e Àrabes nas Margens do Douro', Studium Generale, vol. X, Porto.

Filgueiras, O. L., 1966. O Barco Poveiro, ed. Camara Municipal da Póvoa do Varzim.

Filgueiras, O. L., 1970a. 'Sobre as Origens do Barco Poveiro', Boletim Cultural Póvoa do Varzim, Póvoa do Varzim.

Filgueiras, O. L., 1970b. 'No crepúsculo das embarcacões regionais', in Actas do XXIX Congresso Luso-Espanhol, Colóquio 2—Tome III, Associacão Portuguesa para o Progresso das Ciencias, Lisboa.

Filguerias, O. L., 1977. 'The Xávega-Boat, a case-study on the integration of Archaeological and Ethnological data', in Sources and Techniques in Boat Archaeology, National Maritime Museum, Greenwich. B.A.R. -S29

Gama Barros, H. da, 1947. História da Administracão Pública em Portugal, nos séculos XII a XV, Tomo IV—2nd Edicao, dirigida por Torquato de Sousa Soares, ed. Livraria Sá da Costa/Lisboa.

Garcia y Bellido, A., 1945. España Los Españoles Hace Dos Mil Anos Según La Geografia de Strábon, ed. Espasa-Calpse, Argentina, Buenos Aires, 4th ed.

Greenhill, B., 1976. Archaeology of the Boat, Adam and Charles Black, London.

F. Guillén, J., 1969. La gran família de los trincados y otras embarcaciones regionales con influencias normandas, Instituto Histórico de Marina, Madrid.

Hasslöf, O., 1972. 'Main principles in the Technology of Ship-Building', in Ships and Shipyards, Sailors and Fishermen, Rosenkilde and Bagger, Copenhagen.

Hornell, J., 1970. Water Transport, Origin and Early Evolution, David & Charles, Newton Abbot, 2nd ed.

Ibarra Grasso, Dick and Julio, 1949. História de la Navegación Primitiva, Francisco Liverali—Editor, Buenos Aires.

Kinsey, W. M., 1829. Portugal illustrated, by the Rev., B. D. fellow of Trinity College, Oxford and Chaplain to the Right Hon. Lord Auckland, Published for the author by Trentel and Würtz, Trentelzun and Richter, London, 2nd ed.

McKee, E., 1976. 'Identification of timbers from old ships of north-western European origin', in The International Journal of Nautical Archaeology and Underwater Exploration, vol. 5, no. 1, London.

Marsden, 1976. P., 'A boat of the Roman period found at Bruges, Belgium, in 1899, and related types', in The International Journal of Nautical Archaeology and Underwater Exploration, London.

Marsden, P., 1977. 'Celtic Ships of Europe', in Sources and Techniques in Boat Archaeology, National Maritime Museum, Greenwich. B.A.R. -S29

Matthes, W., 1940. 'The Suevi' in Vorgeschicht der Deutschen Stämme, dir. Reinerth (Hans), Berlin.

Mattoso, J., 1968. Le Monachisme Ibérique et Cluny—Les Monastères du Diocèse de Porto de l'an mille à 1200, Publications Universitaires de Louvain.

Mollat, M., 1964. 'Notes sur La Vie Maritime En Galice Au XII Siècle D'Après L'Historia Compostellana'. Anuário de los Estudios Medievales, Instituto de História Medieval de España. vol. I, Barcelona.

Moore, A., 1925. Last Days of Mast & Sail, David & Charles, Newton Abbot (reprinted 1970).

Nascimento, A. A., 1973. 'Alguns vocábulos portugueses num manuscrito latino do século XV', in Portugalia Historica—I, Lisboa.

Novaes, M. P. de., 1913, Anacrisis Historial, Biblioteca Pública Municipal, IV—vol. II, Porto.

Pico, M. A. T. C., 1963. Terminologia Naval Portuguesa Anterior a 1460, ed. Soc. de Língua Portuguesa, Lisboa.

Ribeiro, O., 1955. 'Portugal', vol. V da Geografia de España Y Portugal, Manuel de Terán, ed. Montaner Y Simón, S. A., Barcelona, 1st. ed.

Ribeiro, O., 1975. 'Portugal, Formacão de' in vol. III, Dicionário de História de Portugal, Joel Serrao.

Ribeiro, O., 1977. Introducões Geográficas à História de Portugal—Estudo crítico. Imprensa Nacional—Casa da Moeda, Lisboa.

Sampaio, A., 1923A, 'As Póvoas Marítimas do Norte de Portugal', in Estudos Históricos e Económicos, ed. Livraria Chardron, vol. I, Porto.

Sampaio, A., 1923B. 'O Norte Marítimo', in Estudos Históricos e Económicos, vol. I, ed. da Livraria Chardron de Lello, Irmao, Lda. Porto.

Schulten, A., 1959/1963. Geografia Y Etnografia Antiguas de la Peninsula Ibérica, ed. Consejo Superior de Investigaciones Científicas, Madrid.

Serrão, J., undated. 'Transportes' in Dicionário de História de Portugal, Joel Serrão.

Smolarek, P., 1969. Studies on Gdansk Pomerian Boat-Building in Xth-XIIIth Centuries. Gdansk Maritime Museum's Proceedings, Gdansk.

Soares, T. de S., undated. Anotacoes à ob. cit. Gama Barros (Henrique).

Viterbo, Fr. J. R., 1798/9. Elucidário, ed. critica de Mário Fiúza, Livraria Civilizacao Editora, Porto. (Reprinted 1966.)

Woolley, Sir L., 1929. Ur of the Chaldees, Penguin Books, London. (Reprinted 1950.)

KEYWORDS

20th century; Post-medieval; medieval.

Portugal; R. Douro; R. Lima; R. Minho.

Clinker-built boats; carvel-built boats.

Boatbuilding characteristics; shell technique; settlement; fishing; river-borne trade.

Model	Function	Area of operation	Average dimensions (m)	Particulars	Propulsion	Crew	Fig.
Rabelo	cargo ⩾ 40 casks	Upstream Entre-os Rios	19.50 x 4.40 x 1.70	Boat bow. Stern castle. Steering oar.	Square sail. 4 oars. poles. sirga.	10-12	3 4
Barquinha rabela (Little Rabelo barge)	cargo ⩽ 40 casks	id.		id.	id.	8-10	3
Barca de passagem (ferry)	ferry	id.	13.00 x 3.30 x 0.50	Boat (or rabão) bow. Steering-oar	2 oars. (cable)	1-2	3
Rabao de apegadas (Rabao with stern castle)	cargo	id.	16.70 x 3.10 x 1.40	Rabão bow. stern castle. steering oar.	Square sail. 4 oars. poles. sirga.	6-8	3 5 8
Rabão de rabo baixo (Rabẽo with low tail)	cargo	id.	13.00 x 3.00 x 0.80	Rabão bow. steering oar.	id.	4-6	6
Rabãozinho (Little rabão)	cargo ferry fishing	id.	9.00 x 2.00 x 1.00	id.	id.	2-3	6 7 8, 26
Rabão Branco (White Rabao)	cargo	ebb-tide Entre-os Rios		beak bow.	sprit-sail. 2-3 oars. poles.	2-3	

Table 1 (contd.)

Model	Function	Area of operation	Average dimensions	Particulars	Propolsion	Crew	Fig.
Rabão carvoeiro (Rabão for coal transport)	cargo	Concas-Porto	23.10 x 5.50 x 2.00	id.	4 oars. poles.	2-5	8 9 10
Barco de frete (cargo)	cargo	Arnelas Casa Branca (near Porto)	17.00 x 3.45 x 1.20	id.	double sprit-sail. 4 oars. poles.	2-4	11
Barco das padeiras (Bakers-women boat)	cargo ferry	Avintes-Porto	7.80 x 1.80 x 0.75	id.	sprit-sail 2 oars.	1-2	
Valboeiro	cargo ferry. fishing	id.	7.30 x 1.75 x 0.67	id.	id.	1-2	13 14
Barco da areia (boat for sand transport)	cargo	Cantareira (mar Foz do Douro)	15.00 x 2.90 x 1.10	id.	id.	2-4	12 26

Table 1 Traditional types of R. Douro boats

Model	Function	Area of Operation		Particulars	Propulsion	Crew	Figs.
Barca (barge)	cargo ferry	Between Viana and Ponte do Lima	16.00 x 3.50 x 0.70	little transom aft. Rudder	square sail or sprit-sail. Poles	2	20 21 26
Barquinho (Little boat)	cargo ferry fishing	id.	10.00 x 2.18 x 0.60	Large transom aft.	sprit sail. Poles	2	22 23
Barca de passagem (ferry)	ferry	Porto da Barca	8.50 x 3.20 x 0.50	transom fore and aft.	poles and cables.	2	
Barco	ferry fishing	Upstream Ponta de Lima	4.30 x 1.40 x 0.38	large transom aft.			

Table 2 Traditional types of R. Lima boat

Model	Function	Area of operation	Dimensions (m)	Particulars	Propulsion	Crew	Figs.
Carocho (bug)	cargo fishing	ebb-tide Moncão	4.15 x 1.60 x 0.45	sheerline	lug sail* oars. poles.	2–4	24 26
Barco (boat)	cargo ferry fishing	Moncão	id.	sheer aft	oars. poles	2	
Barco da sacada	fishing	Lanhelas	7.25 x 1.92 x 0.56		lug sail oars poles	2–4	25

* '...one of the most ancient of sails (...) they came into general use in Europe later than all those hitherto described'. (A. Moore, 206).

Table 3 Traditional typesof R. Minho boat

North Portugal			Galiza/Spain	
R. Douro	R. Lima	R. Minho	Arosa, Pontevedra, el Grove, Sta. Maria de Lanzada	El Ferrol, La Coruña
Rabelos & Valboeiros	Barcos & Barquinhas	Carochos	Dorna	Trincado
keeless flat bottom	keeless flat bottom	T keel (intermediate type)	Keel cf. Fig. p. 78 (top) A. E. Christensen Tr. 1968	Central plank as a keel cf. Fig. p. 78 below A. E. Christensen Tr. 1968
block stems	stem (rabbeted) + transom aft	stem & stern without rabbets	stem & transom aft	stem & stern
single skin	single skin	single skin	single skin	single skin
lap: bevelled for receiving the over-lap	lap: normal	lap: bevelled for receiving the over-lap	lap: bevelled for receiving the over-lap	
	lapped corner		V Transition between bottom planks + keel Fig. p. 78 (top) A.E. Christensen Jr. 1968	L planks
nails hooked over so as to re-enter the planks	nails hooked over so as to re-enter the planks	nails hooked over so as to re-enter the planks		
caulking: fibres driven in after assembly	caulking: fribes driven in after assembly	caulking: fibres driven in after assembly		
framing: long and short work alternating, scarfs and rising used	asymmetrical timbers/ alternating pairs cf. the German <u>dreibordnachen</u> (Neckar)	ribs: floor timber + futtocks	ribs: floortimber + futtocks	ribs: floor timber + futtocks
ribs: angle of incidence for and aft		attenuated angle of incidence fore and aft		

Table 4 (contd.

R. Douro	R. Lima	R. Minho	Arosa, Pontevedra, el Grove, Sta. Maria de Lanzada (Galiza)	El Ferrol/Coruña (Galiza)
cross section: cf. enlarged round-hull Bridgewater barge (A. E. Christensen Jr. 1977 Fig. 184) ≠ Bruges boat (Jonckheere, in Marsden Fig. 5) not derived from dug-outs but from planked canoes	cf. Somerset hull turf boat & East German Baltic coast boat (Greenhill Figs. 3 & 128) & the German <u>binnenboot</u> & probably Falsterbo pram Blomquist, in Smolarek Fig. 31) ≠ Zwammerdam boat (in Marsden Fig. 29): no L planks; not derived from dug-outs but from tubs	round-hull cf. Graveney boat (Green- (Greenhill Fig. 155) & Oselver boat (A. E. Christensen Jr. 1977 Fig. 156)	cf. Fig. p. 78 (top) A. E. Christensen Jr. 1968	
rowlocks: transitional type of <u>Keip</u> + Dalarma type + other variations of Nordic types; the oars rowing free	no oars	normal rowlocks with <u>tholepins</u>	transitional type of <u>Keip</u>; the oars rowing free	normal rowlocks with <u>tholepins</u>
square sail sprit sail	square sail: wooden truss on the square rig cf. (Smolarek Fig. 100; sprit sail.	lug sail	lug sail	two lug sails

Table 4 Comparative data for the traditional N. W. Peninsular boats

5. THE MEDIEVAL SHIPS OF LONDON

Peter Marsden

Summary

London's significance as a chief port for the past nine hundred years is reflected in the discovery of ships and boats there, not only as wrecks, but also as re-used timbers. Some examples of these are illustrated here and briefly set into their historical context.

In recent years several shipwrecks of medieval and later date have been found in the River Thames, together with fragments of other ships and boats that had been reused in the building of the early waterfronts of London. The dated ancient ships, boats and parts of vessels that have been found in central London are listed below. The sites are numbered on Figure 5.1.

ROMAN

1.	Blackfriars ship 1	- shipwreck	- 2nd century
2.	New Guy's House boat	- abandoned vessel	- 2nd century
3.	County Hall ship	- abandoned vessel	- 3rd century
4.	Iron anchor, Blackfriars	- broken mooring	- probably Roman

SAXON PERIOD

5.	New Fresh Wharf boat	- reused planking and rib	- 8-9th centuries

MEDIEVAL PERIOD

6.	Blackfriars ship 3	- shipwreck	- 15th century
7.	Blackfriars ship 4	- shipwreck	- 15th century
8.	Blackfriars boat 5	- reused rib	- 15th century
9.	Custom House boat 1	- reused planking	- 13th century
10.	Custom House boat 2	- reused rib	- 13th century
11.	Trig Lane boat	- reused planking	- 14th century
12.	Bridwell boat	- reused planking	- 15th century

POST-MEDIEVAL PERIOD

13.	Blackfriars ship 2	- shipwreck	- 17th century

More than five other vessels are said to have been found in central London, two of the Roman period, one of the Viking period, and two of the 15th century; while two 18th century Thames barges were found several years

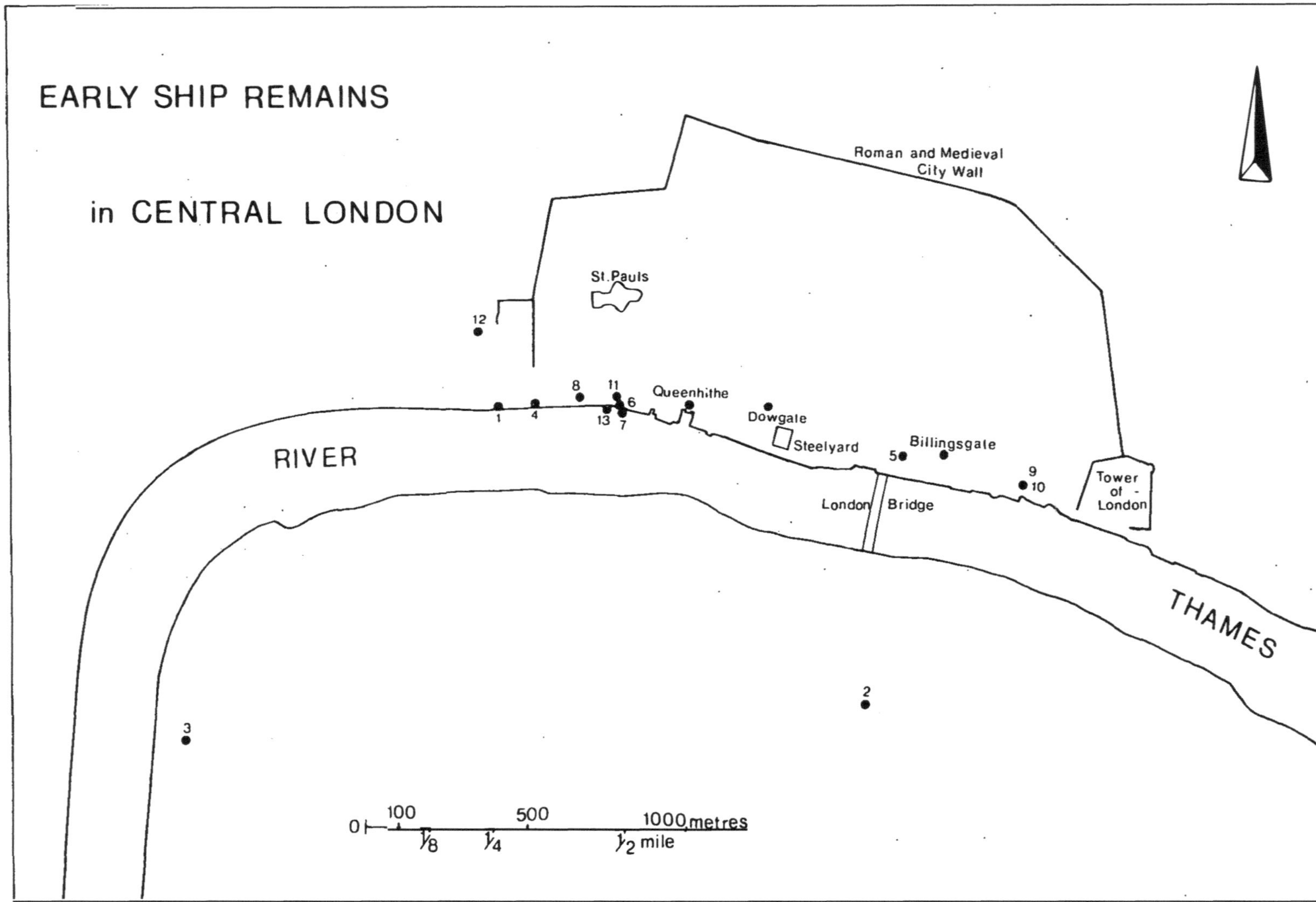
EARLY SHIP REMAINS
in CENTRAL LONDON
Roman and Medieval City Wall
St.Pauls
Queenhithe
Dowgate
Steelyard
Billingsgate
Tower of London
London Bridge
RIVER
THAMES
1
2
3
4
5
6
7
8
9
10
11
12
13
0
100
500
1000 metres
1/8
1/4
1/2 mile

Fig. 5.1

Fig. 5.2 Fragments of a boat of the 8th-9th centuries from the New Fresh Wharf site in London

ago at Battersea reused to form the sides of a dock. The dating of all but the barges at Battersea should be considered as uncertain, and their identification as ships suspect. In spite of this, however, these early discoveries do indicate how important London is as a site for early ships of many periods.

Of the recent post-Roman discoveries the following are particularly important:-

The New Fresh Wharf boat (Fig.5.2) is dated by Carbon 14 to the 8th or 9th centuries (Birm-548 and Harwell-1422). It is one of the most interesting early vessels to have been found in London since it does not have the type of construction that is expected of north-western European boats at that period. It was clinker built of oak, and had a luting of moss. Instead of the overlapping planks being held together by the usual iron rivets, the plank fastenings were of small wooden pegs expanded by small wooden wedges. This construction is only found at that period in ships on the Baltic coast of Poland, where it is believed they were the product of a local Slavic tradition of shipbuilding. We are soon to make an analysis of the moss luting for traces of pollen which may reflect the vegetation of the surrounding country side where the boat was presumably built, and give us a clue to where this might have been. We know so little about ships in Britain at this time, however, that it is also possible that it was a local vessel.

The boat was found in a broken-up condition, the fragments having been dumped on the sloping foreshore, together with other material, probably to form a landing place where vessels could be pulled ashore on the waterfront of Saxon London. This may have been the site of a Saxon landing place, perhaps where a waterfront market existed of the type which certainly existed at Queenhithe in the late 9th century.

This is not the only beach waterfront site found in London, however, for a little further upstream a mud beach was discovered some years ago at Dowgate where, in the 12th century, cargoes were apparently off-loaded onto the shore. In this case the shore-line was found littered with Germanic pottery, including a considerable quantity of Pingsdorf ware. This site lies adjacent to the 'Steelyard', which from at least the late twelfth century was the headquarters in London of the Hanse merchants; but as German merchants are known to have been in London since about 1000 A.D. it is likely that their occupation of the Steelyard site began somewhat earlier.

The ships and boats of the medieval period found in London, were all clinker built, though few have yet been completely studied, Most, however, seem to represent the local river and estuarine traffic rather than the larger seagoing sailing ships, though this view will be tested as far as possible by an analysis of the pollen that may be preserved in the luting of the plank overlaps.

Sufficient of the Custom House boat 1 (Fig.5.3) has been found for us partly to reconstruct the type of vessel, even though the craft had been broken up and re-used in a late 13th or 14th century waterfront. It was a fairly small double-ended clinker built vessel with oak planking. Its flat bottom had been broken into large slabs and reused, while the curving sides, including probably the gunwale, were broken into smaller pieces before reuse.

Fig. 5.3 Parts of a 13th century boat re-used in a waterfront on the Custom House site in London

Fig. 5.4 Blackfriars ship 3, a 15th century sailing barge. The ribs and mast-step had been removed during excavation

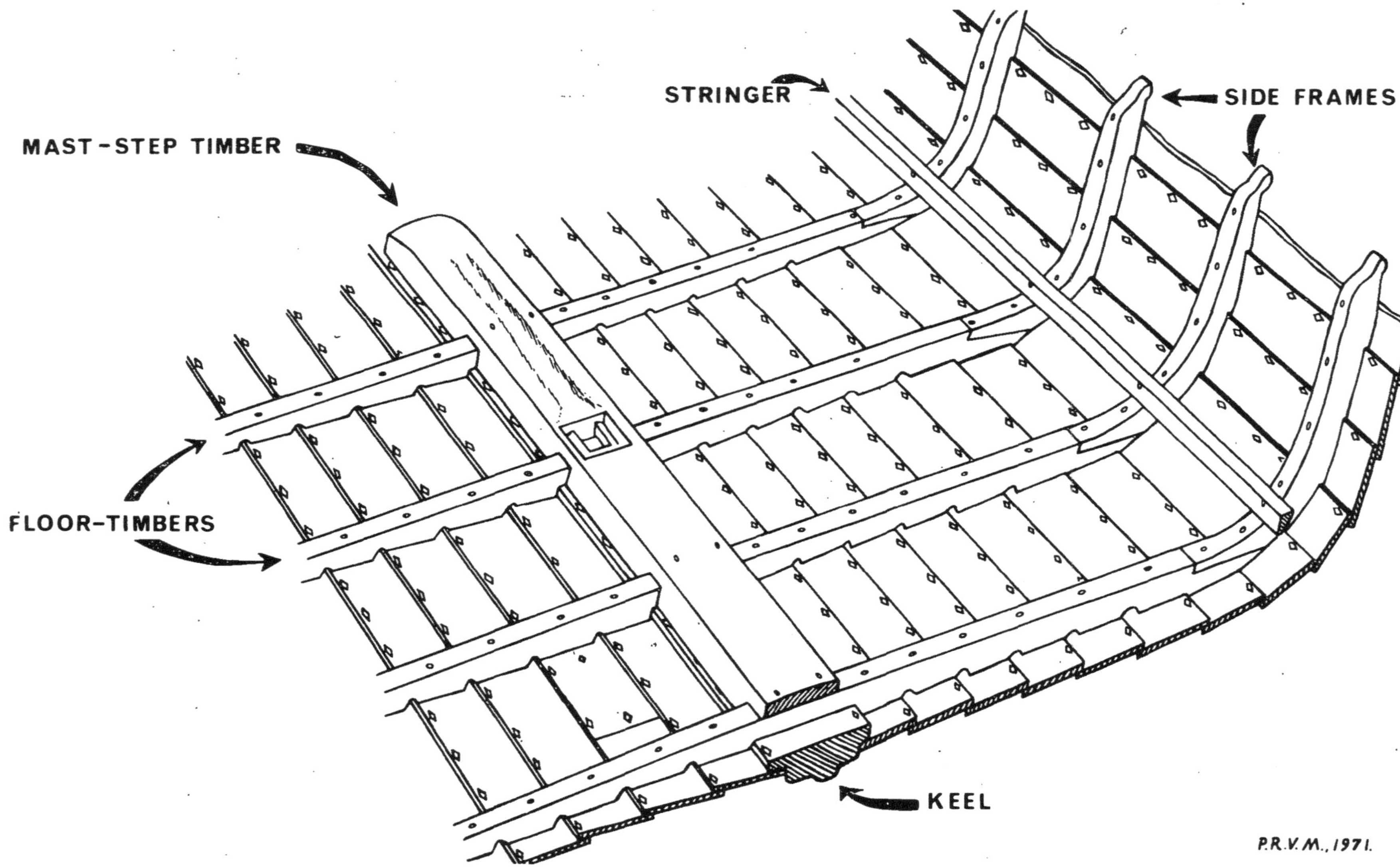

Fig. 5.5 Diagram to illustrate the construction of Blackfriars ship 3

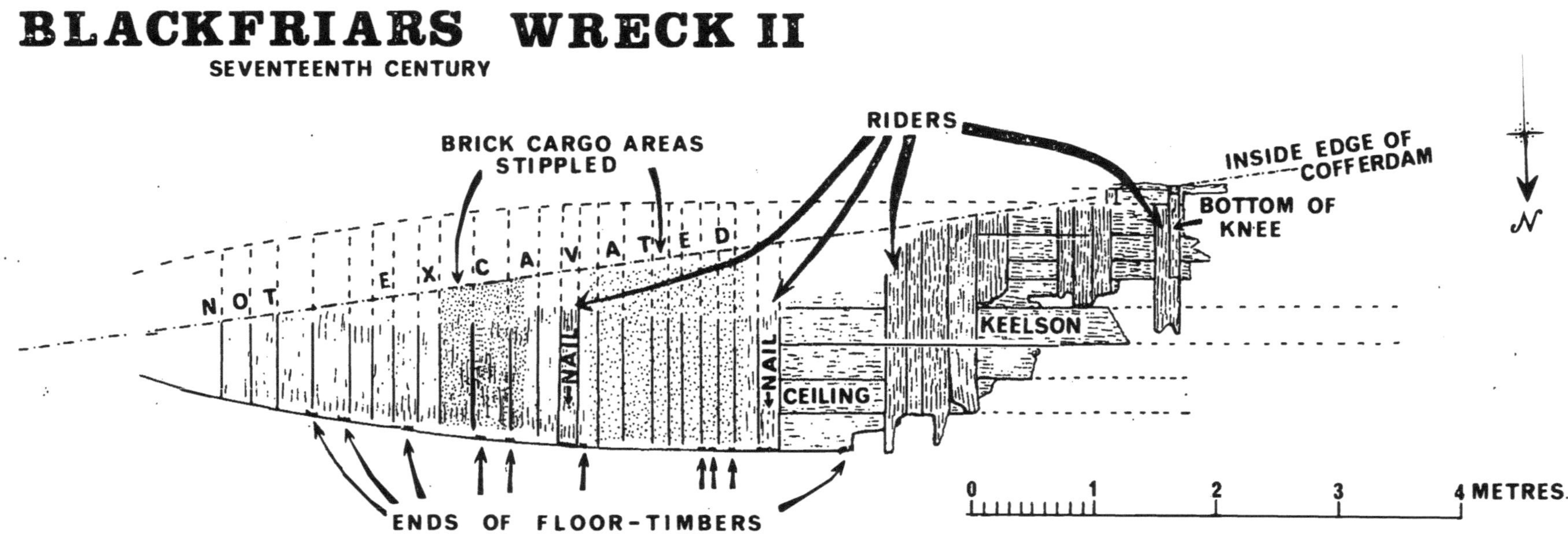

Fig. 5.6 Blackfriars ship 2, a wreck of the 17th century carrying a cargo of bricks

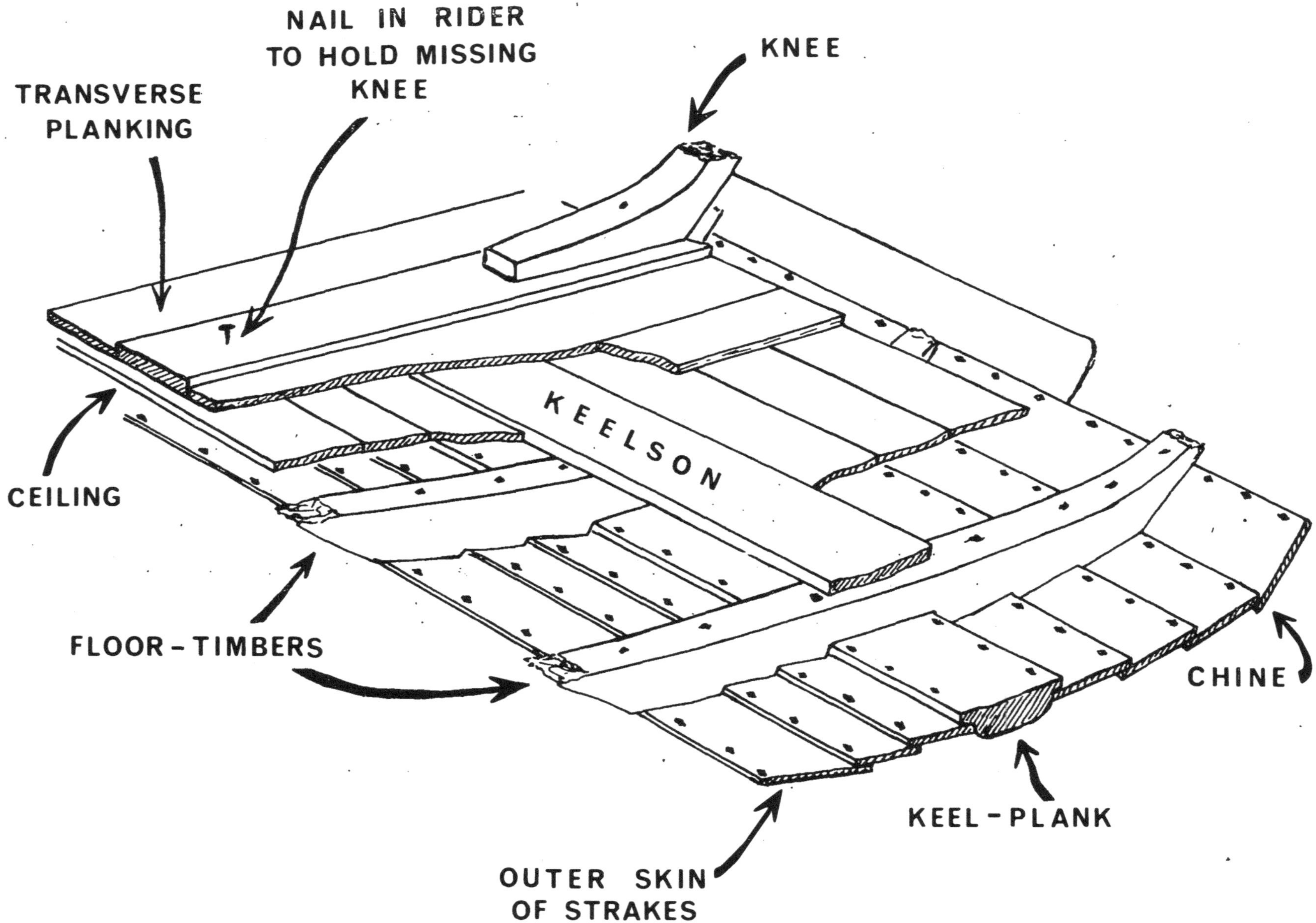

Fig. 5.7 Diagram to illustrate the construction of Blackfriars ship 2

Two of the later medieval wrecks found at Blackfriars had been sunk by accident during the 15th century. Blackfriars ship 3 (Figs.5.4-5), a clinker built sailing barge of oak was pointed at both ends. Its mast could be lowered, presumably to pass under London Bridge. It contained no cargo, but instead a considerable amount of pottery and two fine pewter pilgrim badges that had probably been washed into the vessel after sinking. There were also more than a thousand lead weights from a 15th century fishing net that had been caught on the wreck. Blackfriars ship 4 was only partly recorded, but it appeared to be a somewhat similar clinker-built vessel which had been sunk with a cargo of building stone, quarried near Maidstone in Kent. Its last voyage, therefore, was down the Medway river to the Thames estuary, and then upstream to London.

The seventeenth century wreck, Blackfriars ship 2 (Figs.5.6-7), was of interest, not only because it was carrying a cargo of bricks perhaps for re-building London after the 'Great Fire' of 1666, but also because it too was clinker build, and its method of construction was still essentially that of the Middle Ages (Marsden 1971).

When research is completed, a major publication is planned as a study of all the London finds of Roman, Dark Age, Medieval and later ships and their fittings, in relation to the trade goods, contemporary waterfronts and historical records relating to London.

FURTHER READING

1. Roman

Marsden, P. R. V., 1966. A Roman ship from Blackfriars, London Guildhall Museum Publication.

Marsden, P. R. V., 1965, 'A boat of the Roman period discovered on the site of New Guy's House, Bermondsey, 1958', Trans London & Middlesex Archaeological Society, 21:118-131.

Marsden, P. R. V., 1974, 'The County Hall Ship', IJNA, 3:55-65.

2. Saxon Period

Marsden, P. R. V., 1975, Medieval Archaeology, 19:224.

Marsden, P. R. V., 1976, Medieval Archaeology, 20:166.

3. Marsden, P. R. V., 1977, 'Blackfriars wreck 3. A preliminary note' IJNA, 1:130-132.
See also: Medieval Archaeology, 15 (1971) p. 176; Medieval Archaeology, 18(1974) p. 202; Post-Medieval Archaeology, 10 (1976) p. 163-164.

4. Post-Medieval Period

Marsden, P. R. V., 1971, 'A 17th century boat found in London' Post-Medieval Archaeology, 5:88-98.

5. <u>Other finds</u>

See: <u>Mariner's Mirror</u>, 49:144-5.

KEYWORDS

Medieval.

Britain; River Thames; London.

Boats and ships; revetments; beaches; waterfronts.

Hanse merchants; clinker boatbuilding; Polish boatbuilding; pollen analysis

6. MEDIEVAL LOGBOATS OF THE RIVER MERSEY—A CLASSIFICATION STUDY

Sean McGrail and Roy Switsur

Summary

Samples from nine logboats, recovered from the River Mersey or nearby between 1889 and 1971, have been dated by radiocarbon to the 9th to 12th century a.d. It is argued that a probable in-use date in the 11th century a.d. is indicated. The boats' attributes are analysed and certain features are provisionally identified as common to the group, enabling a description to be given of the characteristic techniques used by 11th century a.d. logboat builders of the River Mersey. It is suggested that this analytical method may be extended to planked boats and ships so that any regional and temporal variations may be identified within a general type.

1. INTRODUCTION

In Logboats of England and Wales a method of describing logboats was outlined which used 115 attributes, each with two or more states (McGrail, 1978A:307-8, 350-8). It was further suggested (1978A:140-2, 331-7) that until numerical taxonomy was shown to be practicable, logboats could best be summarily described using a set of eight general variables:

1. Date (Attributes 15 to 18)
2. Region (Attributes 19 and 20)
3. Function (Attribute 109)
4. Form (Attribute 86)
5. Method of propulsion (Attribute 110)
6. Speed potential (Attribute 111)
7. Size (Attributes 3-5, 115)
8. Boatbuilding characteristics (Attributes 6-14, 21-74)

At the time of writing Logboats (mid 1977) so few boats had been given a reliable date that it seemed impracticable to attempt to classify by boatbuilding characteristics for, although these could be listed for some of the boats, such a scheme would have little meaning without consideration of the time dimension.

In late 1977 and in 1978 further dates became available from the Godwin Laboratory at Cambridge, including dates for eight of the logboats known to have been found in or near the River Mersey and its tributary the River Irwell (Figs. 6.1 and 6.2). Thirteen logboats or fragments are known to have been recovered from this area: Barton (Figs. 6.3 and 6.4) found in 1889 and Irlam (Figs. 6.5 and 6.6) found in 1890, during preparatory work for the Manchester

Fig. 6.1 Distribution of logboat finds in England and Wales. The numbers refer to Catalogue entries in <u>Logboats of England and Wales</u>. National Maritime Museum

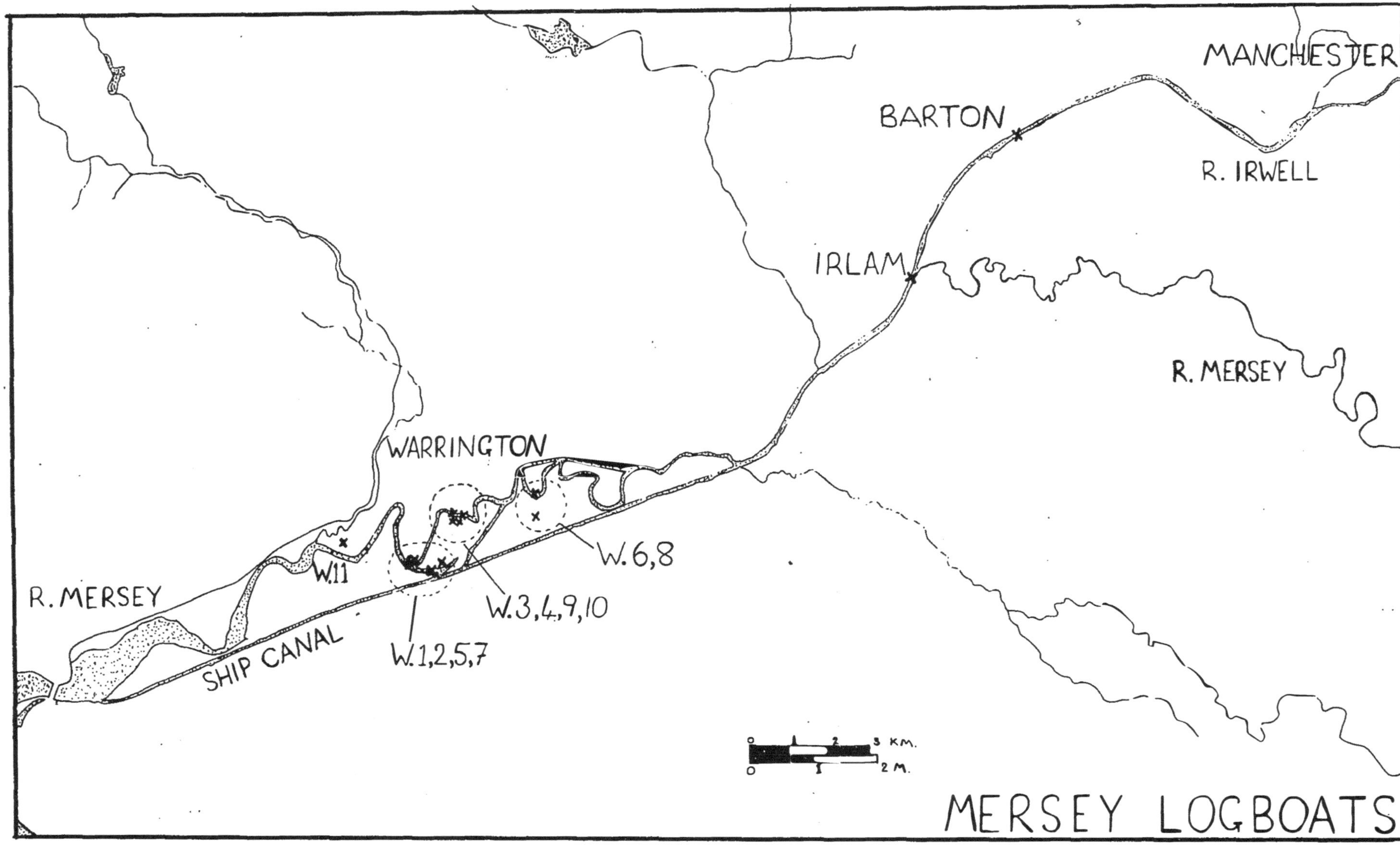

Fig. 6.2 The present course of the River Mersey and River Irwell between Warrington and Manchester. National Maritime Museum

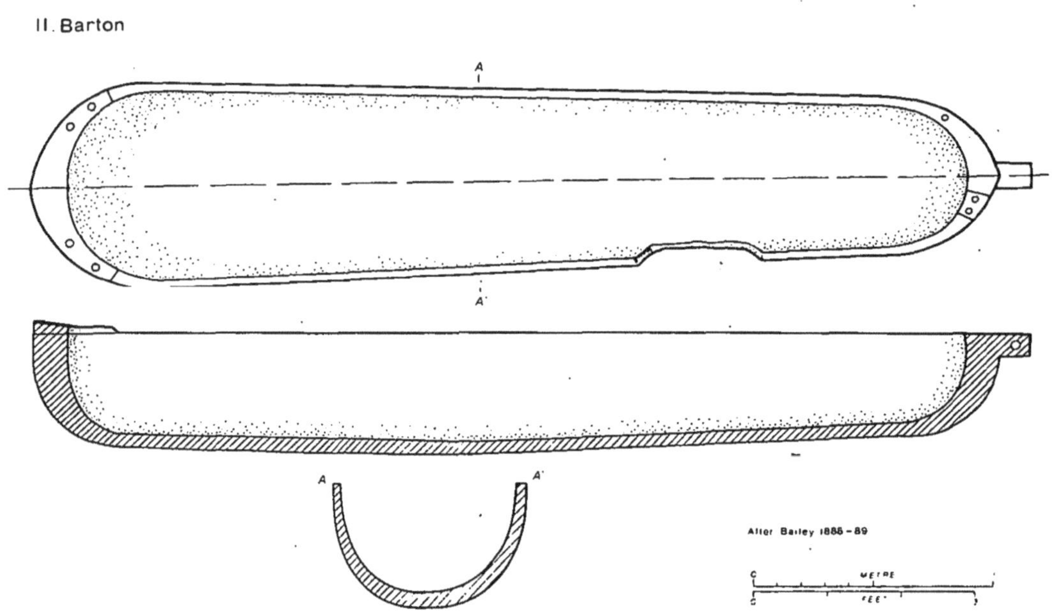

Fig. 6.3 The remains of the Barton logboat as recorded by Bailey in 1888-9. National Maritime Museum

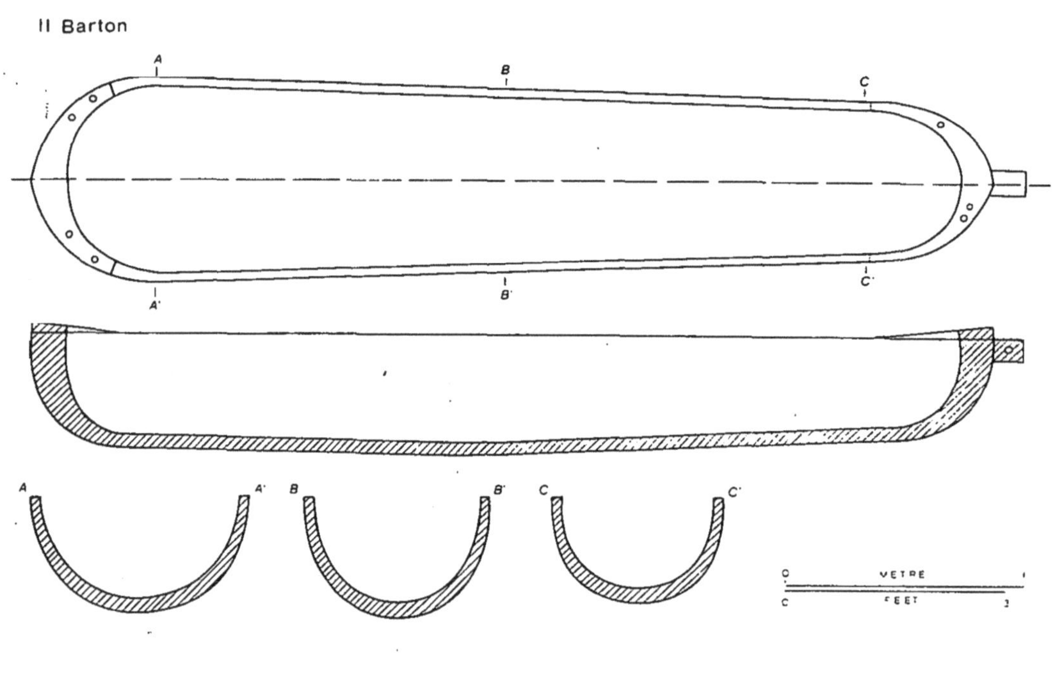

Fig. 6.4 The conjectural reconstruction of the Barton logboat. National Maritime Museum

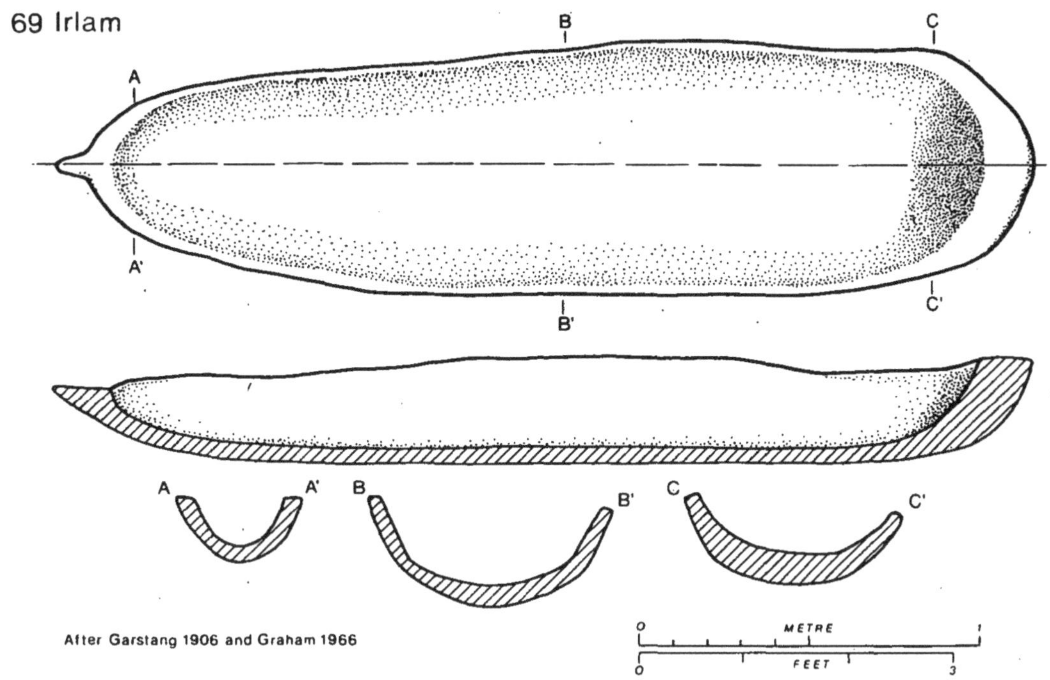

Fig. 6.5 The remains of the Irlam logboat as recorded by Graham in 1966, with the addition of the beak recorded by Garstang in 1906. National Maritime Museum

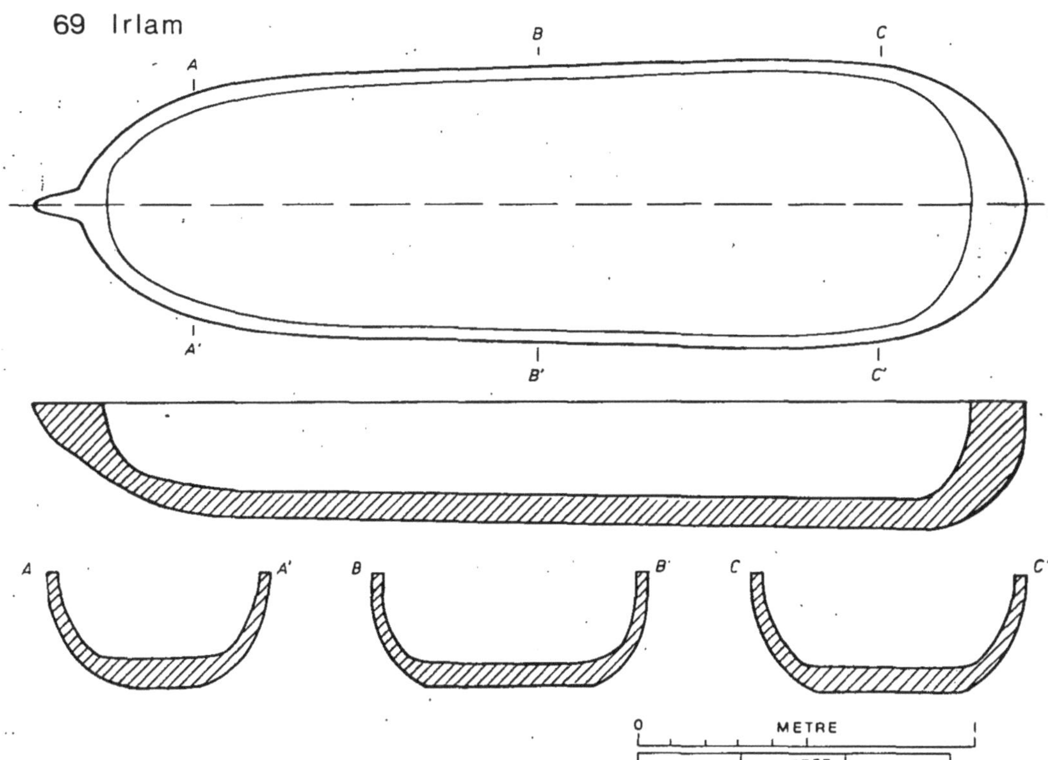

Fig. 6.6 The conjectural reconstruction of the Irlam logboat, including an allowance for the shrinkage experienced before it was recorded. National Maritime Museum

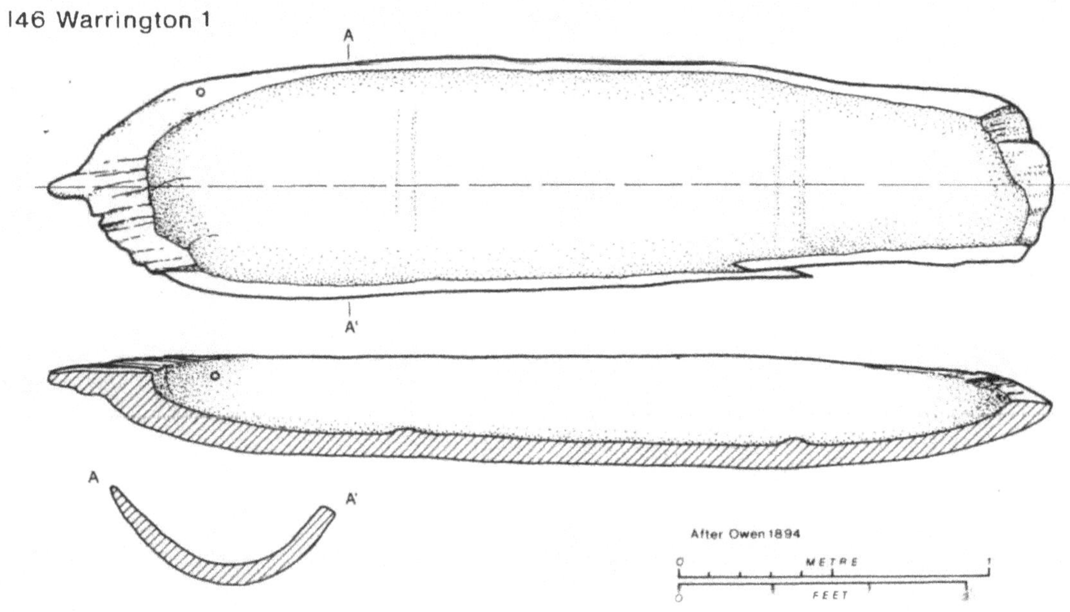

Fig. 6.7 The remains of Warrington 1 logboat as recorded by Owen in 1894, with additional details noted by McGrail in 1974. National Maritime Museum

Fig. 6.8 The remains of Warrington 11 logboat photographed in 1971. Warrington Museum

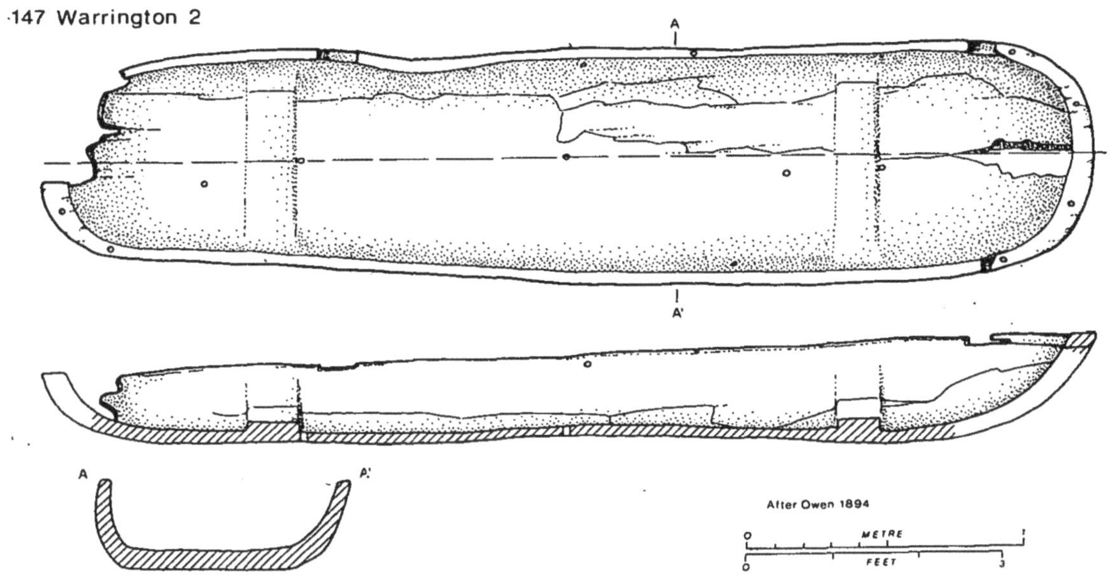

Fig. 6.9 The remains of Warrington 2 logboat as recorded by Owen in 1894, with additional details noted by McGrail in 1974. National Maritime Museum

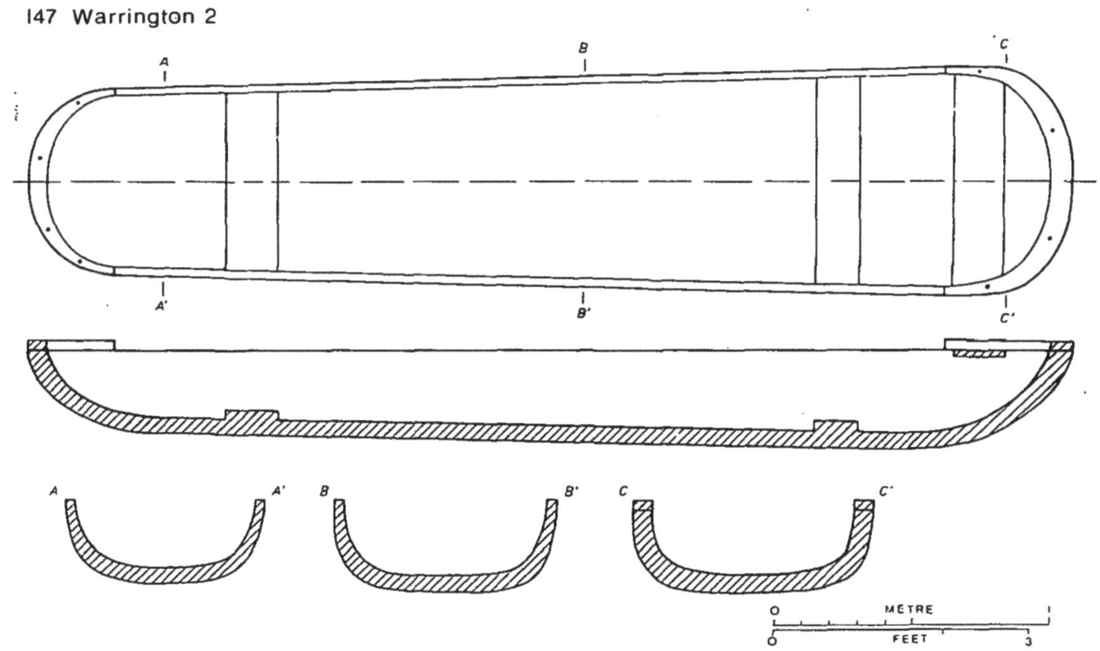

Fig. 6.10 The conjectural reconstruction of Warrington 2 logboat. National Maritime Museum

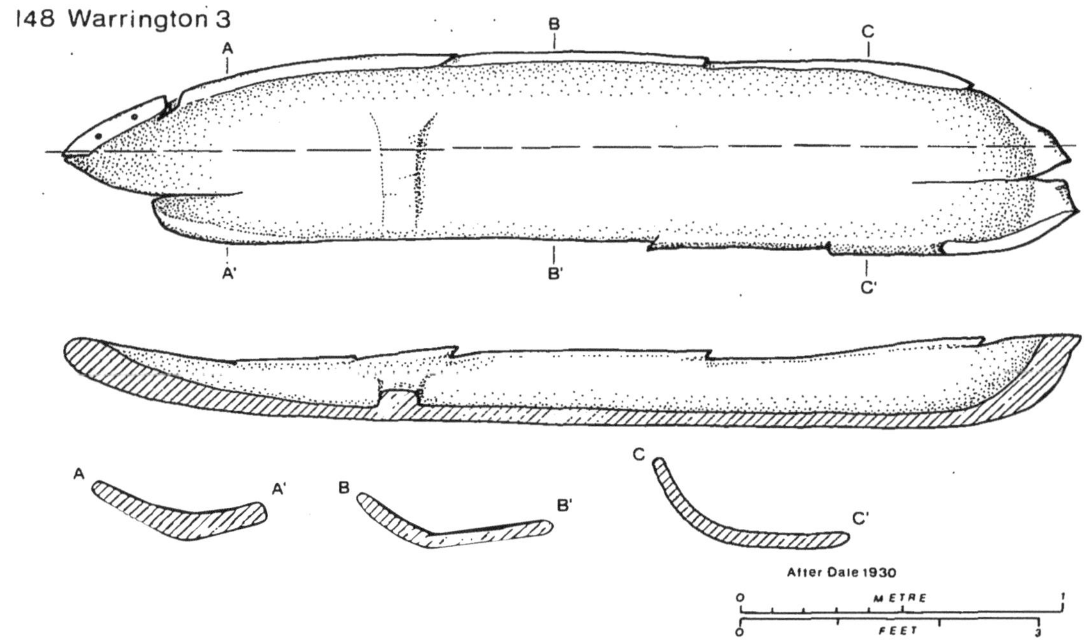

Fig. 6.11 The remains of Warrington 3 logboat as recorded by Dale in 1930, with additional details noted by McGrail in 1974. National Maritime Museum

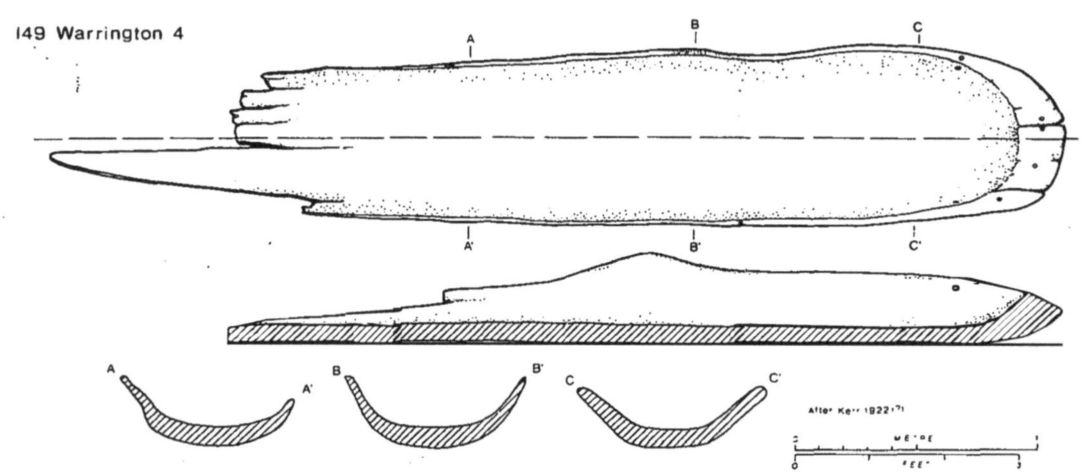

Fig. 6.12 The remains of Warrington 4 logboat as recorded by Kerr in (?)1922, with additional details noted by McGrail in 1974. National Maritime Museum

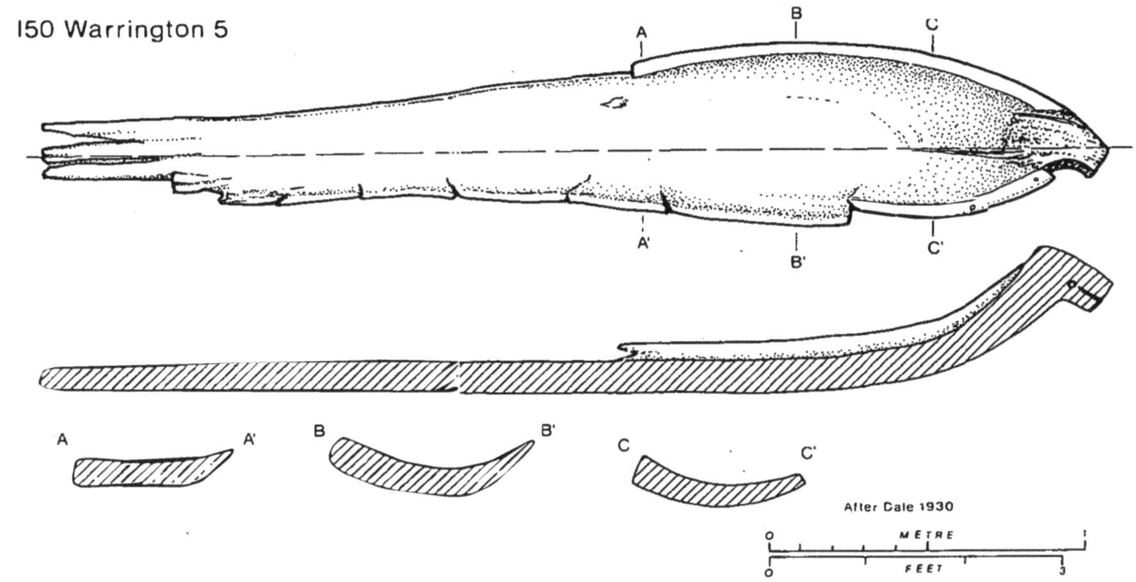

Fig. 6.13 The remains of Warrington 5 logboat as recorded by Dale in 1930, with additional details noted by McGrail in 1974. National Maritime Museum

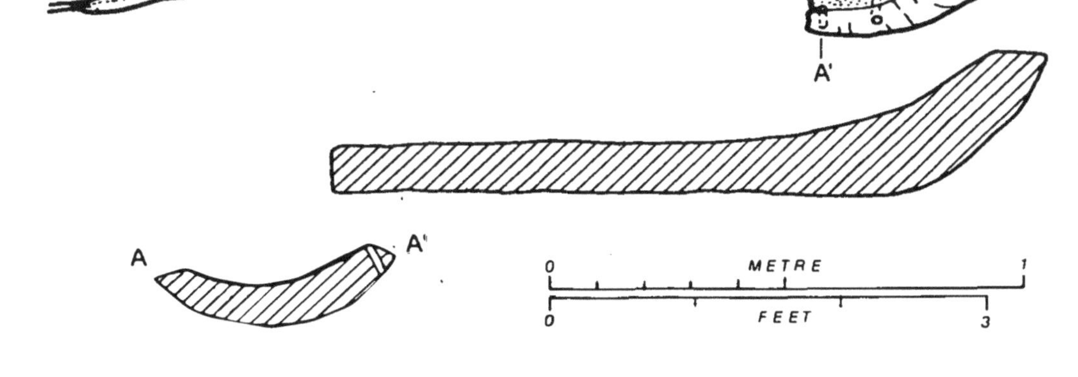

Fig. 6.14 A fragment of Warrington 7 logboat recorded by McGrail in 1974. National Maritime Museum

Ship Canal; and eleven boats (Figs. 6.7 to 6.14) found between 1893 and 1971 during industrial excavation, dredging and other riverine works at Warrington. It is probable that these boats were deposited not far from where they were used and built (McGrail, 1978A:330).

Thus detailed investigation of shared features, especially boatbuilding techniques, becomes practicable for such a group of boats from a restricted area, with many of them dated.

The remains from Barton are now in the Manchester University Museum albeit in a somewhat inaccessible position, and those from Irlam are in the Salford Museum. Warrington boats 1 to 8 are in the reserve collection of Warrington Museum, and boat 11, an end fragment only, is under conservation there. Warrington 6 and 8 were not drawn when originally found and are now very fragmentary with few significant features. Warrington 9 and 10 were only noted as finds and there are no surviving remains. Details of these boats are given in the appropriate catalogue in McGrail (1978A:Ch.9). For reasons there explained it has proved possible to make hypothetical reconstructions of only Barton, Irlam, and Warrington 2. Thus only limited analysis is possible for the other boats.

There are in fact three other finds from the Mersey drainage area (Astbury, Cholmondley, and Oakmere) but as these are from lake sites near the source of rivers feeding the Mersey, rather than river sites, these will not be considered further here.

2. DATING

The Mersey logboats had been subjected to various conservation measures during the years since their recovery. Some had been allowed to dry naturally but others had been treated. The substances used in their preservation included compounds containing carbon which originated from widely differing age ranges, and so posed considerable problems in the purification stage of the dating procedure. Some of the materials were of modern origin and so relatively highly radioactive, whilst others were so ancient as to contain but negligible amounts of radioactive carbon. The samples sent to the Godwin Laboratory, Cambridge were subjected to a thorough series of purification measures in order to remove this contaminating carbon and so isolate the wood of the logboat itself. Only when this purity was assured was the measuring process carried out. The wood was oxidised in high pressure pure oxygen and converted to carbon dioxide. This was purified further in order to remove all traces of electronegative impurities such as the oxides of sulphur and nitrogen as well as any halogens and the water vapour. The radioactivity was measured in a gas proportional counter at about three atmospheres pressure until some 40,000 disintegrations had been timed. (Switsur, Hall and West, 1970; Switsur and West, 1973). The counting rates were compared with those from the international oxalic acid dating standard and the instrumental background, corrected for changes due to variable temperature and barometric pressure. The errors in these measurements were combined in quadrature and the final standard deviation expressed as an uncertainty of the date in years.

The eight recently obtained dates are listed chronologically in Table 1 together with a radiocarbon date published earlier for Warrington 1 (Shotton

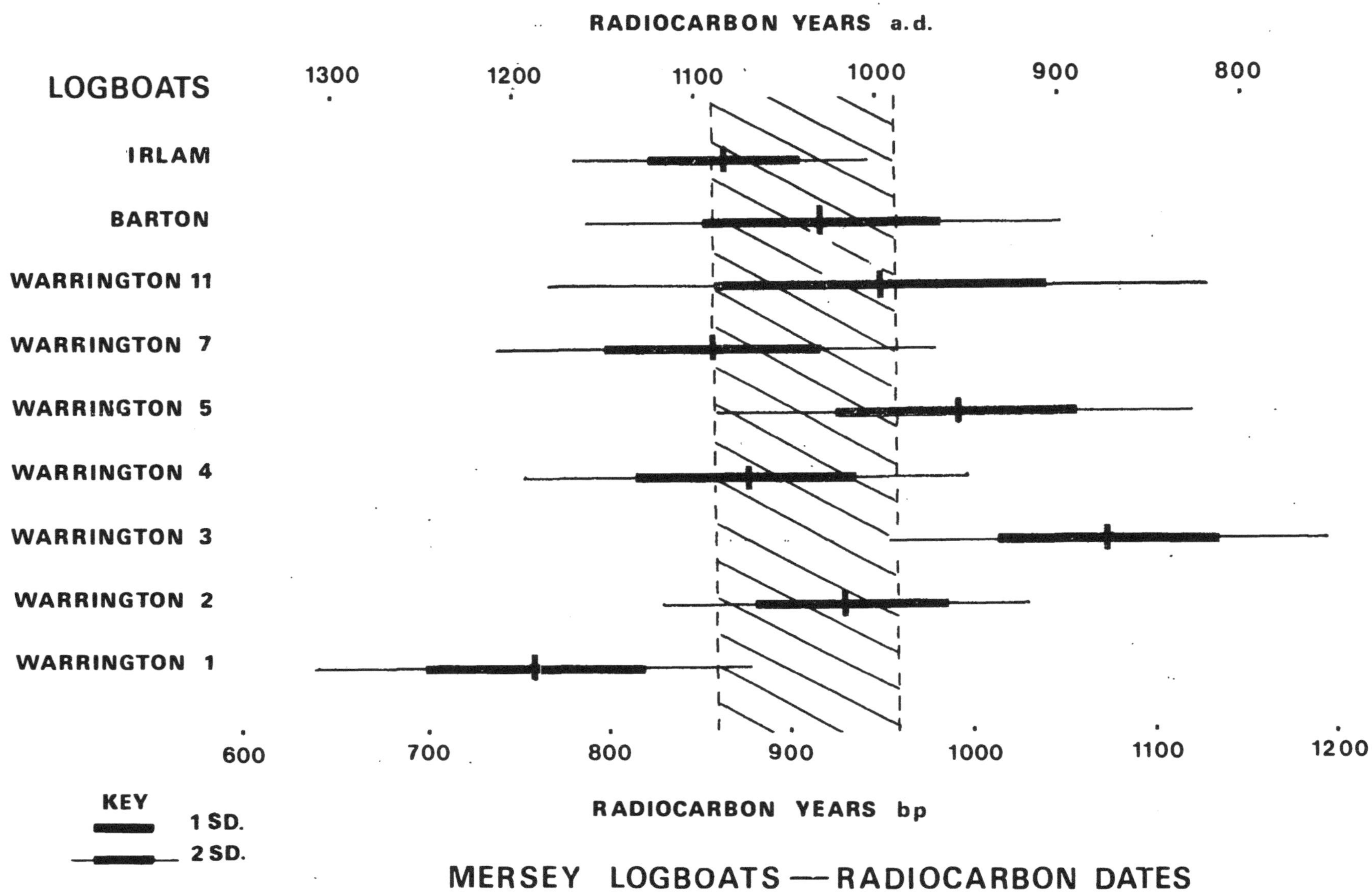
RADIOCARBON YEARS a.d.
1300
1200
1100
1000
900
800
LOGBOATS
IRLAM
BARTON
WARRINGTON 11
WARRINGTON 7
WARRINGTON 5
WARRINGTON 4
WARRINGTON 3
WARRINGTON 2
WARRINGTON 1
600
700
800
900
1000
1100
1200
RADIOCARBON YEARS bp
KEY
1 SD.
2 SD.
MERSEY LOGBOATS — RADIOCARBON DATES

Fig. 6.15

and Williams, 1973). These are conventional radiocarbon dates based on the Libby Half-life for radiocarbon of 5568 years and the uncertainty (±) term represents one standard deviation of the counting statistics.

It should be noted that the convention used is the so-called 'English System' in which the lower case letters ad, bc, bp are used for conventional radiocarbon dates; upper case letters AD, BC being used for calendar years. At the radiocarbon dating congress of 1976 in California it was agreed that Dating Scientists would express radiocarbon dates solely as B.P. (i.e. with respect to the calendar year 1950) and this is how the definitive dates will appear in the journal Radiocarbon. It is hoped that the problems associated with publishing dates in two different ways will be resolved at a future International conference.

The samples for radiocarbon dating, except that from Warrington 11, were collected during 1974/75 by McGrail from logboats in the Manchester University, Salford and Warrington Museums, and were taken from a position on the surviving remains identified as nearest to the outside of the original log in order to get as close to the felling date as is now possible. It is estimated that for Barton, Irlam, and Warrington 1, 2, 4, 7, these samples were within 30-50 rings of the final growth ring. The remains of Warrington 3 and 5 are now less extensive than the others, and suitable samples could only be obtained from positions estimated to be within 100-125 rings of the final growth ring. Warrington 6 and Warrington 8 remains are even scantier and, as they have few if any significant features, samples from these were not dated. There are no known remains of Warrington 9 and Warrington 10.

Individual radiocarbon dating laboratories have distinctive procedures, and emphasis is placed differently at the various stages. It is instructive to study the prefaces to the papers published by each laboratory in the journal Radiocarbon to note the many and various ways the results are calculated and the uncertainty terms established. Due to this non-standardisation, only rarely are dates from more than one laboratory suitable for detailed study and correlation (Switsur, forthcoming). In this present work all dates except one are from a single laboratory and thus detailed study of their inter-relationships becomes possible.

It will be seen from Table 6.1 and Fig. 6.15 that the dates form a relatively close grouping in time, lying between about 1075 and 760 bp (a.d. 875 and a.d. 1190) with some 300 years separating the mean dates. The central dates for six of the boats, those for Irlam, Barton, and Warrington 2, 4, 7 and 11, all lie within the period 860 to 960 bp (a.d. 1090 to 990) and the dates for Warrington 1, 3 and 5 enter this range within the uncertainty band of two standard deviations. As noted previously, the outermost parts of Warrington 3 and 5 have not survived so that samples for these boats could not be obtained from a relative position comparable with the other boats. An allowance for the effect of this discrepancy in sampling could well bring the central dates for these two boats also within the 860 to 960 bp period. Thus the dates for the outermost used rings (approximating to the log-felling dates) of six of these nine boats are indistinguishable, and eight possibly lie in the period 860 to 960 bp, that is approximately in the 11th century a.d. The outermost used rings in Warrington 1, however, appear to date to the 12th or even 13th century

Laboratory Reference	McGrail's Cat. No.	Logboat	Conventional Radiocarbon Date		
			bp	ad	±
Q-1392	148	Warrington 3 (Howley)	1075	875	60
Q-1394	150	Warrington 5 (Arpley)	992	958	65
Birm-269	156	Warrington 11 (Gate Warth)	950	1000	90
Q-1391	147	Warrington 2 (Walton Lock)	930	1020	50
Q-1396	11	Barton	920	1030	65
Q-1393	149	Warrington 4 (Fairclough)	878	1072	60
Q-1456	69	Irlam	865	1085	40
Q-1395	152	Warrington 7 (Walton Arches)	860	1090	60
Q-1390	146	Warrington 1 (Arpley Meadow)	760	1190	60

Table 6.1 Radiocarbon Dates for Logboats from the River Mersey

Note: Samples were oak (*Quercus* sp) except that from Warrington 11 which was elm (*Ulmus* sp)

a.d., although it is possible that they too may be 11th century. These dates are *termini post quem* for their use as boats. There is ethnographic evidence that (partly-worked) logs were sometimes stored underwater for several years before being fashioned into logboats and this may occasionally have been the case in antiquity (McGrail, 1978A:33-4). However, it seems likely that these Mersey logboats were in use within 25 years of the felling date and possibly within one year.

In order to give historical dates A.D., comparable with the precision now possible in some regions and eras from dendrochronology, radiocarbon dates must be calibrated; but there is at present no general agreement which calibration curve or table to use. An attempt has been made to calibrate these nine dates for the Mersey boats: using curves published by Clark (1975) and by Stuiver and Suess (1976). The detailed argument will be published elsewhere (Switsur, forthcoming), but using these methods it seems likely that the calendar dates for Warrington 2, 4, 7, 11, Irlam and Barton lie in the range A.D. 1060 to A.D. 1205, that is generally in the 12th century A.D.

Thus, using either "raw" or calibrated radiocarbon dates, albeit only a single date for each boat, it can be seen that 8 of the 13 logboats recovered from a stretch of river only 12 miles (20 km) in length, were built from logs which were probably felled within the short span of 100 to 150 years. Warrington 1, on present indications, may be somewhat outside this range but sufficiently close as to be considered a member of the group. The four undated boats have insignificant or no remains and contribute only a little to the study of characteristic features. It thus seems entirely justifiable to examine the data from these Mersey boats to see if there are any common features and in particular whether there is any recognisable pattern in the detailed methods of boatbuilding, a concept which Crumlin-Pedersen (1977:164) has described as: "a certain local practice at a given time and place".

3. GROUP CHARACTERISTICS

(a) Function

It is probable that medieval lgoboats were generally used for ferrying, fishing, fowling and the collection of reeds. However, as a by-product of investigations into transverse stability, it can be shown that logboats have significantly differing load-bearing capabilities which may be equated with different functions, and this concept can be used to compare them (McGrail, 1978A:141). Two boats (Irlam, Warrington 2) are 1st-rate bulky-cargo carriers and Warrington 2 is equally good at carrying personnel. Barton on the other hand is assessed as a 2nd-rate high density cargo carrier, but there are reservations about the accuracy of the early drawings of this boat (McGrail, 1978A:158) and a future examination when the remains are more accessible, may result in the postulation of different reconstruction, with consequences on the capabilities of the boat. The variability of this and other measures of performance means that Function, in the sense used here, cannot be considered a group characteristic.

(b) Form

Barton, Irlam and Warrington 2 all taper in plan and elevation and have half-rounded transverse sections, thus conforming to the shape of the log (McGrail, 1978A:99, 322): for the remainder there is insufficient evidence. Four of the boats (Barton, Irlam, Warrington 1, Warrington 2) are of canoe form as defined in McGrail (1978A:130); and five others (Warrington 3, 4, 5, 7, 11) are of provisional canoe form: that is, probably canoe form, but insufficient remains or records to be certain.

(c) Propulsion

There is no surviving evidence for the use of oars of sail in these boats thus the use of paddle, or in certain circumstances pole, is assumed (McGrail, 1978A:320-1).

(d) Speed potential

Barton and Warrington 2 are assessed as having 'good' paddle-power and 'medium' speed potential, resulting in 'good speeds'. Irlam has 'moderate/good' paddle-power and 'slow' potential speed resulting in 'average speed'. This is too variable to be considered a common characteristic (McGrail, 1978A: 141-2, 335-6).

(e) Size

Barton, Irlam, Warrington 2 and probably Warrington 1 and 3 are short, being in the 1st or 2nd of 6 groups for length (see Attribute 4 in McGrail, 1978A:350); nevertheless they are in the range of lengths (2.77 m–4.65 m) within which 50% of the analysable boats lie (McGrail 1978A:325-6). The reconstructed maximum beams for the three boats which can be assessed vary from 0.85 m (Irlam) to 0.91 m (Barton) and are thus within the range of breadths (0.73 to 0.99 m) within which 67% of assessable English and Welsh boats lie (McGrail, 1978A:326).

The deadweight capabilities at standard freeboard of the three boats which can be assessed range from 85 kg (Irlam) to 658 kg (Barton), and are in the 1st, 2nd and 4th groups into which Attribute 115 has been partitioned (McGrail, 1978A:357-8). This is too variable to be considered common to the group.

(f) Boatbuilding characteristics

(i) Selection and conversion of timber

The logboats are of oak (*Quercus* sp), except Warrington 11 which was reported to be elm (*Ulmus* sp), and Warrington 8, 9, 10 which have not been identified. Warrington 11 is unique in its species amongst the known logboats of England and Wales.

Barton, Irlam and Warrington 2, and possibly Warrington 1, were made from halves of logs which had been split longitudinally. The relationship of the pith of the log to the sheerline cannot now be identified in the remaining boats. The lengths of the parent logs may be taken to be somewhat longer than the lengths of the boats given in paragraph 3 (e). The maximum girth and volume of the ideal parent log from which the three assessable boats were made are given in Table 6.2.

Logboat	Maximum girth	Volume
Irlam	3.30 m at 0.95 from the surviving butt-end	2.28 m^3
Warrington 2	3.33 m at 0.50 m from the surviving butt-end	2.84 m^3
Barton	4.40 m at 0.50 m from the surviving butt-end	5.53 m^3

Table 6.2 Details of the parent logs of three logboats

The logs used for Warrington 1 and 3 were probably of similar girth and volume. Thus in addition to their short length (paragraph 3 (e), these logs were unexceptional in maximum girth and volume, being amongst the lower values in the data for the 24 English and Welsh reconstructed logboats (McGrail, 1978A:309), where girths range from 3.08 m to 5.96 m, and volumes from 1.85 m^3 to 37.95 m^3.

The percentage of the half-log worked away varied from 88.6% (Barton) to 78.6% (Irlam): these are close to the extremes for the assessable English and Welsh logboats made from half-logs. Moderate, or in the case of Barton moderate/poor, amounts of useable space were obtained when these half-logs were hewn out (McGrail, 1978A:310-1).

(ii) Thickness gauges

Warrington 2 and 8 had thickness gauges (McGrail, 1978A:31) of diameter c. 23-25 mm; Irlam appears not to have any. Barton and Warrington 1 need to be examined for this feature when they become more accessible. The remaining Warrington boats are now fragmentary, and although there is no record of thickness gauges, they may originally have had them, as in the past these were not always detected at first recording with the boat still waterlogged.

(iii) Treenails for fastening fittings

Several of the boats have holes which most probably housed treenails used to fasten fittings to the hull. Such holes in Warrington 1, 2, 3, 4, 7 varied in pre-shrinkage diameter from 23 to 27 mm, although one hole in Warrington 2 was reported by Madeley (1894:101) to have been 30 mm in diameter. Warrington 5 had treenail holes reported to be c. 20 mm in diameter, whilst the diameter of those in Barton was not recorded and cannot yet be re-examined.

(iv) Thickness of bottom and sides.

The pre-shrinkage thicknesses for Barton, Irlam and Warrington 2 were in the range of 100 to 60 mm for the bottom, and 50 to 40 mm for the sides.

(v) Ridges

Warrington 1, 2, 8 each have ridges (McGrail, 1978A:55-6) towards the two ends of the boat though not sufficiently close to have been used as a foot-timber by a man sitting on the ends. Warrington 3 has one ridge; one was not recorded at the other end by Dale (in Dunlop, 1930-1) and the timbers are now too fragmentary to determine whether this feature might have been overlooked. Warrington 6, for which no drawing exists, consisted only of a 1 m fragment which had a thwartship ridge. Where measurable these ridges are c. 100-200 mm wide and c. 40-50 mm deep. Expressed as a percentage of the deduced original length of the boat, the five ridges on Warrington 1, 2, 3 are positioned between 22% and 36% from their respective ends (cf. McGrail, 1978A:315).

(vi) Thwarts

The thwart associated with Warrington 2 no longer survives; it was reported to be a 20 mm thick plank, c. 0.18 m in breadth, which evidently fitted into notches in the boat's top edge, held in place by two of the vertical treenails which also passed through the crescent shaped transverse-strengthening timber around the end (see (vii) below). A man sitting on this thwart could have used the nearby ridge as a foot-timber. There is some evidence that Warrington 1 also had a thwart at one end.

(vii) Transverse strengthening fittings

Barton and Warrington 2 had natural crooks fastened by vertical treenails around the curve of flat-topped ends (McGrail, 1978A:65-6; 318-9). Although there is only direct evidence for the fitting of these timbers at one end—probably the stern— of these two boats, it is likely that there were similar fittings at the bow ends. Vertical holes around the ends of Warrington 1, 3, 4 suggest that similar crooks were fastened there. Except for Irlam which appears to have no trace of this form of strengthening fitting, all other boats from the Mersey left insufficient remains for the former presence or absence of this fitting to be established.

(viii) Miscellaneous fittings

Barton, Warrington 5, 11, 1, 7 and Irlam have, or are reported to have had, beaks—that is horizontal projections from the upper portion of the bows (McGrail, 1978A:68). The first three boats had horizontal holes through these beaks, probably for a painter; the other three beaks also formerly may have had holes. The surviving remains of the other boats in this group are such that they could all have had a beak in the now lost end fragment.

4. DATE RANGE FOR BOATBUILDING TECHNIQUES

Table 11.8 in McGrail (1978A:332-3) gives the earliest and latest dates for the appearance of certain boatbuilding techniques in securely dated English and Welsh logboats. Three of these ranges may now be extended and the feature "thwarts" added (see Table 6.3).

Feature (Attribute No.)	Earliest appearance	Latest appearance
Half-rounded transverse section (81)	1500 b.c. ± 150 Chapel Flat Dyke (possibly) 295 b.c. ± 50 Poole	a.d. 1190 ± 60 Warrington 1
Ridges (47,48)	834 b.c. ± 100 Brigg	a.d. 1190 ± 60 Warrington 1
Beak with horizontal hole (67)	a.d. 814 ± 60 Llyn Llangorse	a.d. 1190 ± 60 Warrington 1
Thwart (56)	a.d. 1020 ± 50 Warrington 2	a.d. 1190 ± 60 Warrington 1 (possibly)

Table 6.3 Date range of certain boatbuilding features
(Amendments to Table 11.8 in McGrail (1978A)

5. NOTABLE FEATURES ABSENT

In none of these logboats is there evidence which might be interpreted as for fittings to enhance stability i.e. external longitudinal timbers, pairing two boats, or expansion (McGrail, 1978A:38-41, 44-55). Where substantial remains have surrived (Barton, Irlam, Warrington 1, 2, 3, 4) this evidence is just about conclusive.

On these latter boats, and indeed on all the remains, there is also no sign of extension, fitted ribs, fitted transoms, dovetailed or mortice and tenon joints, cleats or bulkheads; nor is there evidence for propulsion by oar or sail.

6. CHARACTERISTICS OF THIS GROUP OF LOGBOATS

The evidence for common characteristics discussed in section 3 of this paper is summarised in Table 6.4.

General Variables	Details	Yes	Probably	Possibly	No	Insufficient evidence
1. Date	11th century a.d.	6	2	1	0	4
2. Region	River Mersey	13	0	0	0	0
4. Form	log-shaped	3	0	0	0	10
	canoe	4	5	0	0	4
5. Propulsion	Paddle/pole	4	0	0	0	9
7. Size	Short 4.17 m	3	2	0	0	8
8. Boatbuilding characteristics	Raw material					
	Oak	9	0	0	1	3
	half-log	3	0	1	0	9
	log volume low	3	2	0	0	8
	Thickness gauges of diameter c. 23-25 mm	2	0	0	1	10
	Treenails for fittings generally of diameter 23-27 mm	7	0	0	1	5
	Ridges	5	0	0	2	6
	Transverse Strengthening fitting at ends	2	3	0	1	7
	Beaks (Note 2)	6	0	0	0	7

Table 6.4 Characteristics held in common by the Mersey logboats

Notes: 1. The numbers in the General Variable column refer to the list of variables given in Section 1 of this paper. Variable 3 (Function) which can only be assessed for those three boats hypothetically reconstructed varies too much to be considered common; Variable 6 (Speed potential) and the deadweight aspect of size (Variable 7) are also too variable.

2. Three of these beaks had horizontal holes; the other three may also.

With due allowance for the large number of entries in the "insufficient evidence" column, we may cautiously propose the following as distinctive of this 11th century a.d. group of Mersey logboats:

> Short oaks with girths in proportion, were selected and split longitudinally. Maximum use was made of these half-logs by giving the boats a half-rounded transverse section and longitudinal taper in elevation and plan. The fashioning of rounded ends thus resulted in a "canoe" form. Beaks with horizontal holes for a painter were left protruding from the bow end. Treenails were used to fasten fittings to the hull, and the holes for these and for thickness gauges were c 23-27 mm in diameter (i.e. "one inch"). Ridges were left across the bottom at c one-third of the length from the ends of the River Mersey boats, but were not found on those from the River Irwell. Transverse strengthening fittings — crooks naturally curved to the required shape — were fastened to the ends of the boats to minimise the risk of the log splitting. There were no fittings for propulsion by oar or by sail.

On the evidence at present available, the above passage describes a group of techniques used by some 11th century a.d. logboat builders on the middle reaches of the River Mersey and its tributary the Irwell. Individual features are not necessarily unique. For example short oak half-logs were used in other regions of England and Wales (e.g. Thornaby, Yorkshire) and comparable methods of strengthening the boats' ends against splitting were possibly used elsewhere (e.g. Shapwick, Somerset). It may be however that this combination of features is diagnostic and defines a regional grouping not found elsewhere in England and Wales at that time. This proposition can only be substantiated or contradicted when further regional groupings—such as those near the Thames; Arun; Tyne, Tees and Wear— are examined, as further radiocarbon dates become available.

7. FUTURE RESEARCH

Dating by dendrochronology, if it proves practicable for this group of boats, should result in dates which are more precise and thus further clarify the discussion of felling dates in part 2 above. Documentary searches and other investigations by local historians and archaeologists may throw light on in-use dates and possibly indicate whether there is any likelihood of a catastrophe having caused a simultaneous deposition of these Mersey boats.

Future dating of other logboats may lead to more precision in the dating of the floruit of the boatbuilding techniques.

It is for consideration that the analytical methods used in this investigation should be used on groups of logboats from other countries. With suitable modifications this combination of attribute analysis and dating might also be applied to planked boats and even to ships. So that, for example, within a group of ships all classified as cogs, changes over time, regional variations and even local differences in building techniques may be recognised. Thus we may see a little more clearly beyond the ships to the shipbuilders.

ACKNOWLEDGEMENT

We wish to thank Miss E.V. Vigeon of the Salford Musuem, Dr. A.J.N.W. Prag of the Manchester Museum, and A. Leigh of the Warrington Museum for allowing access to their collections and for allowing samples to be taken for dating; and Mr. Allen Ward for his expert technical assistance in the Godwin Laboratory, Cambridge.

Figs. 1, 3 to 7, and 9 to 14 were drawn for publication by Eva Wilson from original drawings by Sean McGrail, Fig. 2 was drawn by Anne Yates.

REFERENCES

Clark, R.M., 1975, 'A calibration curve for radiocarbon dates' *Antiquity,* 49:251-66.

Crumlin-Pedersen, O., 1977, 'Some principles for the recording and presentation of ancient boat structures' in McGrail, S., (ed) *Sources and Techniques in Boat Archaeology* : 163-177. N.M.M. Archaeological Series No. 1, Greenwich. B.A.R. (Oxford) S.29.

Dunlop, G.A., 1930-1, 'Three dugout canoes found at Warrington' *Trans. Lancs. and Cheshire Archaeological Society,* 47:16-26.

McGrail, S., 1978A *Logboats of England and Wales.* N.M.M. Archaeological Series No. 2, Greenwich. B.A.R. (Oxford) 51.

McGrail, S., 1978B "A medieval logboat from Giggleswick Tarn, Yorkshire" in Annis, P. (ed) *Ingrid and other Studies.* N.M.M. Monograph 36: 25-46.

Madeley, C., 1894, 'On two ancient boats found near Warrington'. *Trans. Historical Society of Lancs. and Cheshire,* 46:97-106

Shotton, F.W. and Williams, R.E.G., 1973, Birmingham University Radiocarbon Dates VI. *Radiocarbon,* 15:1-12.

Stuiver, M., and Suess, H.E., 1976, 'On the relationship between radiocarbon dates and true sample ages'. *Radiocarbon,* 18:534-40.

Switsur, V.R., Hall, M.A., and West, R.G., 1970, University of Cambridge Natural Radiocarbon Measurements IX. *Radiocarbon,* 12:590-598.

Switsur, V.R., and West, R.G., 1973, University of Cambridge Natural Radiocarbon Measurements XII. *Radiocarbon,* 15:533-544.

APPENDIX 1

PROPOSED REVISION OF THE ATTRIBUTE LIST FOR LOGBOATS

In the light of further experience, it is proposed that the Attribute list given in McGrail (1978A:350-8) should be extended in the following manner:

No.	Attribute	Proposed alteration
34.	Dovetails	This attribute should be divided into three new attributes each with several possible states: the several precise forms of the dovetail defined by drawing; whether it is double or single; and whether it is used as a repair, as a fastening, or as part of a structural member.
52.	Fitted Ribs	This attribute should be given several states to describe the form, as defined in McGrail (1978A: 59). - single piece with double curve - alternating half-ribs plus side-timbers - separate half-ribs - paired alternating half-ribs.
53.	Fitted Transom	This attribute should be given several states to describe the precise form as defined under Stern code 4 in Fig. 205 in McGrail (1978A). (This duplicates to some extent Attribute No. 77 Stern elevation).
72.	Paired logboats	This attribute should be given several possible states, to describe the forms of pairing: Types 1, 2 or 3 as defined in McGrail (1978A:44-7).

As further types of fitting or technique are identified, attributes such as 55 (Thole Pin), 57 (External longitudinal timbers) and 63 (Washstrakes) should also be given appropriate states. It is also possible that in due course new attributes such as 'Rib-spacing' may need to be defined.

In conjunction with Ole Crumlin-Pedersen and Bêat Arnold consideration is being given to the possibility of re-arranging the morphological code (McGrail, 1978A: fig. 205) in a more logical order.

APPENDIX 2

COEFFICIENTS FOR THREE OF THE MERSEY LOGBOATS

In Appendix 4 to Logboats of England and Wales technical aspects of the 24 reconstructed logboats were described by 22 coefficients. One of these coefficients (21) estimates the work done in building the boats, and one (22) estimates the relative proportion of useable space obtained. One (5) assesses the boats' payload at restricted draft and one (20) at minimum freeboard. Eight others (1, 2, 3, 6, 7, 8, 9, 10) were devised to assess the boats' ability to carry men or to carry a range of cargo, under given conditions. Ten others (4, 11, 12, 13, 14, 15, 16, 17, 18, 19) are naval architectural coefficients.

In McGrail (1978B) three further coefficients were defined, which, in conjunction with coefficient 20, assess the logboat at State D (minimum freeboard);

23. Bulky, low density (D) $= \frac{Vcc}{Vl} = \frac{\text{Critical cargo volume } (m^3)}{\text{Volume of log } (m^3)}$

24. Men (D) $= \frac{M}{Vl} = \frac{\text{Maximum number of men carried}}{\text{Volume of log } (m^3)}$

25. Critical Bulk Density (D) That density of cargo which results in zero metacentric height ($\overline{GM_t} = 0$).

Values of these three new coefficients for the three reconstructed Mersey logboats are given in Table 6.5 below; the other 22 coefficients are in McGrail, (1978A:348-9).

Logboat	23. Bulky (D)	24. Men (D)	25. CBD (D)
Barton (11)	0.184	0.904 (including ballast)	772.769
Irlam (69)	0.267	1.754 **	295.268
Warrington 2 (147)	0.352	1.410 (including ballast)	293.519

Table 6.5 Coefficients for three logboats

** one man sitting at the stern (not standing, as is the standard requirement).

By comparing four coefficients at State C standard freeboard (3, 6, 8, 10) with the corresponding coefficients at State D minimum freeboard (24, 25, 20, 23) it can be seen that, with increased draft:

(a) capability of carrying men (coefficient 3 and 24) is improved (except for Warrington 2 which can already hold maximum men at State C).

(b) critical bulk density (coefficient 6 and 25) is increased; thus the ability to carry cargoes of widely different bulk densities is decreased. Barton, which at standard freeboard can carry stone, wheat-grain or turf, can only carry stone or wheat-grain at minimum freeboard.

(c) capability of carrying high density cargo (coefficient 8 and 20) is improved.

(d) capability of carrying bulky low density cargo (coefficient 10 and 23) is worsened for Irlam and Warrington 2, and remains approximately the same for Barton.

Whether these changes alter the relative functions of the three logboats may only be determined when similar coefficients have been calculated for other boats.

KEYWORDS

Medieval.
Britain; River Mersey; River Irwell.
Logboats.
Classification; radiocarbon dating; boatbuilding characteristics; woodworking techniques.

7. PIROGUE MONOXYLE ET PORT FLUVIAL DE PORT-BERTEAU

(Charente-Maritime)

Eric RIETH

Summary

La fouille subaquatique de Port-Berteau, en Charente-Maritime (France), a permis d'étudier l'implantation d'un petit port, en dehors de tout contexte urbain, en fonction uniquement d'un débouché pour des ateliers de production de céramiques médiévales et modernes. Ces céramiques, produites du XIIIè au XVIIIè siècle, étaient destinées en partie à l'exportation via La Charente, vers l'Europe du Nord-Ouest au Moyen-Age et les colonies françaises d'Amérique au XVIIIè siècle principalement. La fouille a permis également d'étudier une embarcation monoxyle, de près de 13 mètres de long, datée de la fin du XIIIè-début XIVè siècle.

Origine de la fouille

Au cours d'une plongée dans la Charente, deux plongeurs de La Rochelle, J.P. FROUIN et D. MAGAL, découvrirent en 1969, sur le fond de la rivière, au lieu dit Port-Berteau (Charente-Maritime) (Fig.1), un dépôt important de céramiques. Cruches, plats, écuelles, en bon état de conservation, s'étendaient sur une surface de plusieurs dizaines de mètres carrés par 7 mètres de profondeur. Cette trouvaille fut l'origine d'un programme de recherches sur la céramique saintongeaise, mené durant 3 ans, de 1971 à 1973.

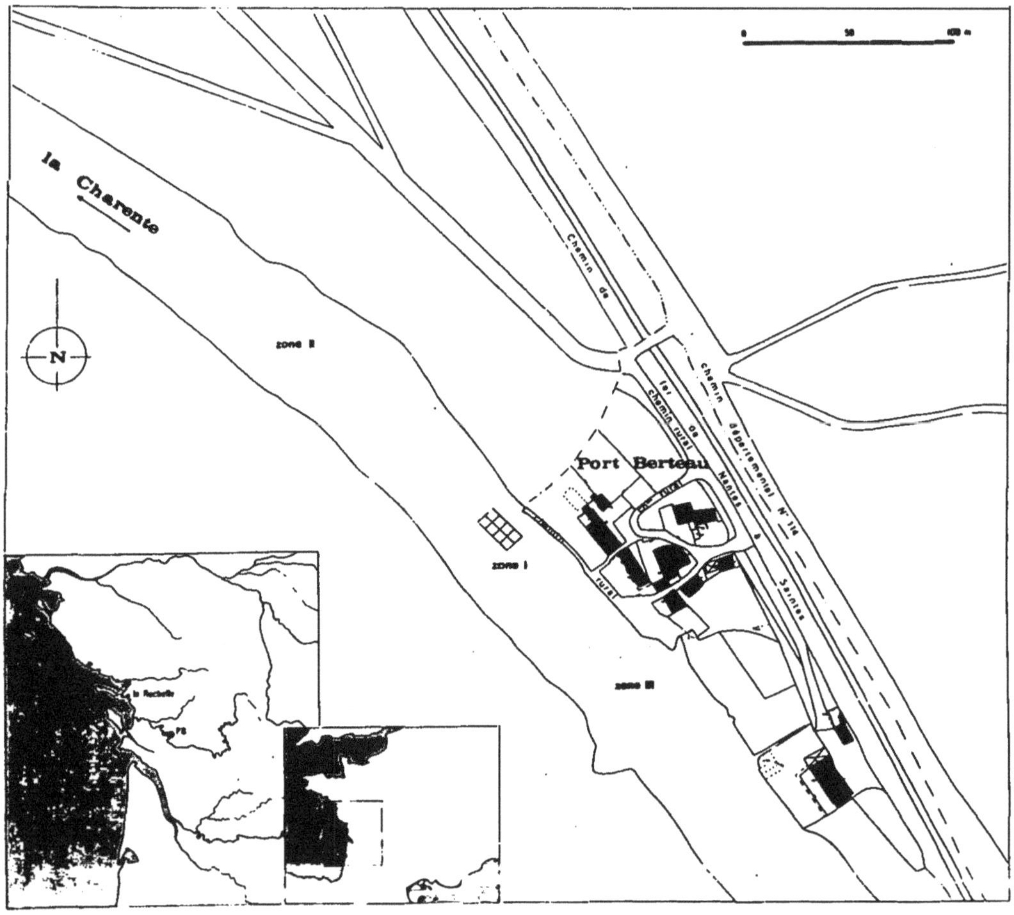

Fig. 7.1 Plan du hameau de Port-Berteau, avec les différentes zones de fouille (Potiers de Saintonge, 1975:109)

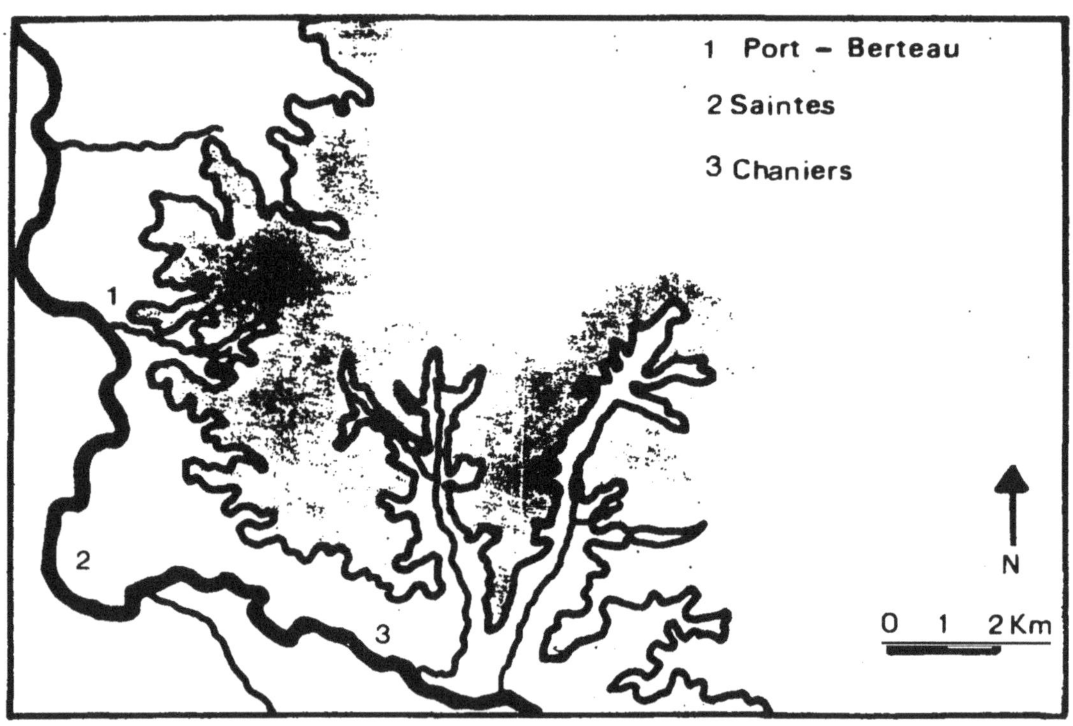

Fig. 7.2 Le site de Port-Berteau; sa relation avec l'arrière pays

La fouille subaquatique de Port-Berteau ne constituait que l'une des approches de cette enquête entreprise par le Musée des Arts et Traditions Populaires de Paris sous la direction générale de Jean CHAPELOT, assistant d'archéologie médiévale à l'Université de Paris I (Potiers de Saintonge..., 1975).

La fouille subaquatique avait pour objectifs l'étude de la structure portuaire, de ses installations annexes, du matériel archéologique en corrélation avec le port, et celle d'éventuelles épaves. Il s'agissait par conséquent d'analyser à partir d'un exemple précis, chronologiquement et géographiquement déterminé, le cadre matériel et les moyens de transport utilisés sur La Charente.
A un autre niveau, la fouille devait permettre d'une part de mettre au point des méthodes de travail subaquatique en rivière et d'autre part de montrer l'intérêt de ce type de recherche en France.

Avant d'aborder l'étude du port et de l'épave monoxyle, il n'est pas inutile de rappeler le paysage et le contexte économique dans lesquels s'intègre Port-Berteau.

Implantation du site portuaire

Le site s'inscrit dans un paysage marqué par deux traits majeurs : La Charente qui, à Port Berteau, subit encore l'effet de la marée et un plateau qui domine la rivière. Revenons un instant sur ces deux points.

Port-Berteau est situé dans le canton de Saintes, sur la commune de Bussac, sur la rive droite de la Charente, à 5 km par la rivière de Saintes. Cette ville portuaire, important carrefour routier par ailleurs, est le point le plus en amont où la marée se fait encore sentir, marquant ainsi la limite du domaine maritime à l'intérieur des terres. Axe de pénétration, la marée constitue aussi un moyen propulsif naturel pour une navigation d'aval vers l'amont. Port-Berteau est, en outre, le dernier port avant Saintes.

Du point de vue du relief, il existe une nette opposition entre les rives gauche et droite (Fig.2). La rive gauche est plate, inondable, constituée d'alluvions fluviatiles. La rive gauche, par contre, est dominée par un plateau de 50 à 70 m. d'altitude. Ce plateau forme un obstacle réel entre la rivière et l'arrière-pays. Or, une faille, partant de Port-Berteau, entaille ce plateau sur plusieurs kilomètres, créant un axe de circulation entre l'arrière-pays et la Charente. Notons que cette situation n'est pas unique ; elle se retrouve par exemple à Chaniers port fluvial en amont de Saintes.

Une question vient immédiatement à l'esprit : pourquoi le port ne s'est-il pas implanté sur la rive gauche, plate, d'accès aisé, largement ouverte sur l'arrière-pays? La réponse se trouve dans le contexte économique de Port-Berteau.

Environnement économique

Le plateau surplombant la rive droite est occupé par une ceinture d'ateliers de potiers. Un texte de 1320 (Potiers de Saintonge..., 1975 : 25) indique "que le chapitre de la cathédrale de Saintes autorise la création d'une chapelle, qui donnera son nom au village de la Chapelle des Pots". On peut considérer que l'installation de ces potiers a du s'effectuer entre 1250 et 1300. A partir de cette date, les ateliers ne s'arrêteront qu'à la

Fig. 7.3 Pichet à motifs décoratifs peints, vers 1250-1350 (Potiers de Saintonge, 1975:61)

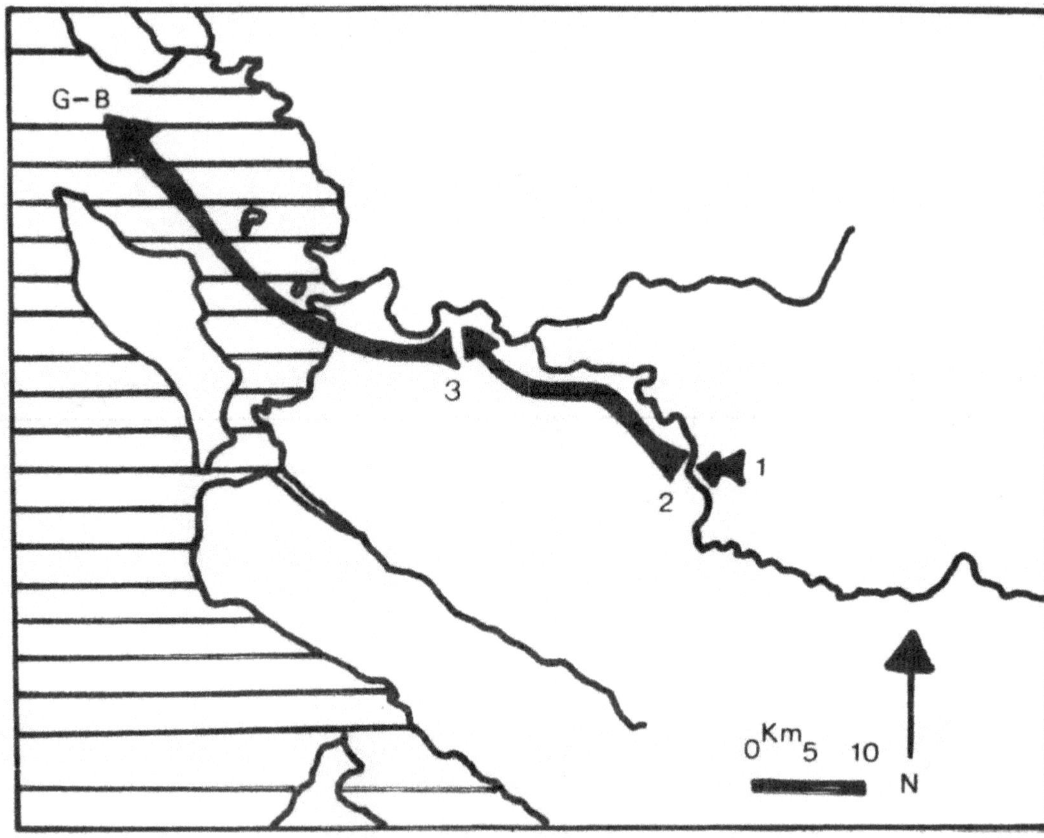

Fig. 7.4 Axe des échanges à longue distance au Moyen-Age au Départ de Port-Berteau

Fig. 7.5 Ecuelle décorée, XVIIIème siècle (Potiers de Saintonge, 1975:85)

fin du XIXè siècle.

Or, les nombreuses céramiques tant médiévales que modernes retrouvées dans la Charente, à Port-Berteau, proviennent de ces ateliers du plateau de Saintes. On constate donc que le choix de l'implantation du port, sur la rive droite, naturellement défavorable, est délibéré et directement lié à la présence de centres de production de céramiques exportées par la rivière. En effet, cette céramique transitant par Port-Berteau est destinée à l'exportation par voie maritime.

Au Moyen-Age, les lieux de trouvailles les plus abondants sont en Angleterre des ports comme Londres, Southampton ou Bristol (Dunning, 1968). Il est évident que cette localisation est à mettre en rapport avec la commercialisation d'un autre produit saintongeais : le vin (Craeybeckx, 1958).A cette boisson de qualité, onéreuse, se greffe une céramique fine (Fig.3) qui se retrouve dans des sites aristocratiques ou ecclésiastiques. Il existe ainsi au Moyen-Age,un axe de circulation privilégié et cohérent entre le vin et la céramique saintongeaise dont Port-Berteau est l'un des points de départ (Fig.4).

A l'époque moderne, et au XVIIIè siècle spécialement, le contexte est très différent. La France possède un vaste empire colonial en Amérique du Nord. Les céramiques produites dans les ateliers du plateau saintois, céramiques de qualité courante (Fig.5) et destinées au milieu rural, transitent par Port-Berteau (Fig.6) à destination de Québec, Louisbourg ou la Louisiane (Dunton, 1971 ; O. et J. Chapelot, 1973).

Il est certain qu'au Moyen-Age aussi bien qu'à l'époque moderne, Port-Berteau ne constituait qu'une étape,

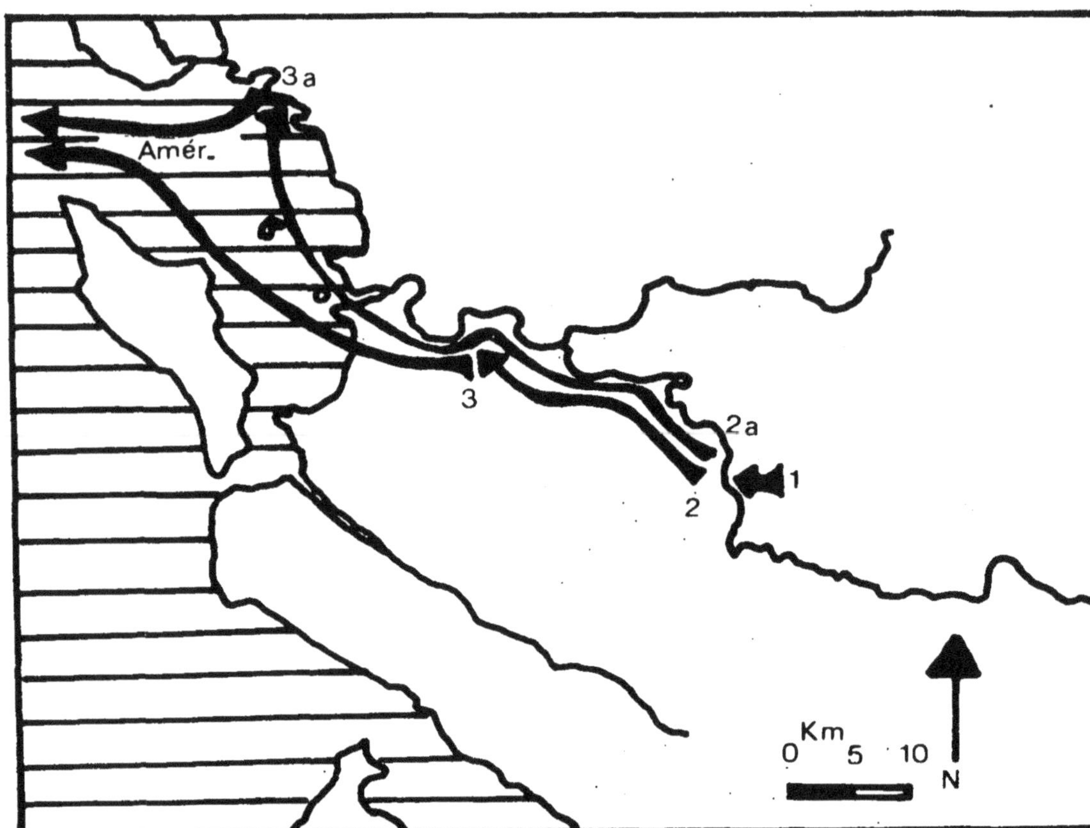

Fig. 7.6 Axe des échanges à longue distance à l'époque moderne au départ de Port-Berteau

Fig. 7.7 Axes des échanges à courte distance au départ de Port-Berteau: rive gauche-rive droite; amontaval

qu'un point d'embarquement strictement fluvial vers des ports proches de l'embouchure de la Charente susceptibles de recevoir des navires de mer, comme Saint Savinien ou Tonnay-Charente. Cet axe de circulation est essentiellement orienté de l'amont vers l'aval. Mais, Port-Berteau est également tourné vers l'amont et s'intègre dans un circuit d'échanges régionaux.

L'axe économique à longue distance amont-aval est recoupé par un axe transversal rive gauche-rive droite (Fig.7) Diverses sources d'archives indiquent l'existence d'un bac (Rieth, 1973 : 109). Celui-ci a fonctionné jusqu'en 1963. Il permettait de gagner la rive gauche et le port de Saintes. Entre outre, différents actes notariés (dont le plus ancien remonte à 1775) (Rieth, 1973 : 110) stipulent que les propriétaires du port étaient aussi cultivateur, meunier ou marchand de bois. On peut, dans ces conditions, raisonnablement penser, sans en avoir une preuve manifeste, que Port-Berteau était également un petit marché rural. Cette double fonction de port et de lieu de transactions n'est pas exceptionnelle en elle-même. Toutefois, il faut bien noter la différence de niveau entre les deux fonctions. L'activité portuaire est, sans aucun doute, dominante dans l'implantation et la vie économique de Port-Berteau. Sans la céramique, produit spécialisé exporté par la Charente, il est fort probable que Port-Berteau n'eut jamais existé.

On constate que cette relation entre un produit spécialisé et un port particulier se retrouve sur d'autres sites de la Charente comme Port-à-Clou, Port la Pierre, Port Tublé (pour port au blé).

Examinons à présent le cadre matériel de ce port.

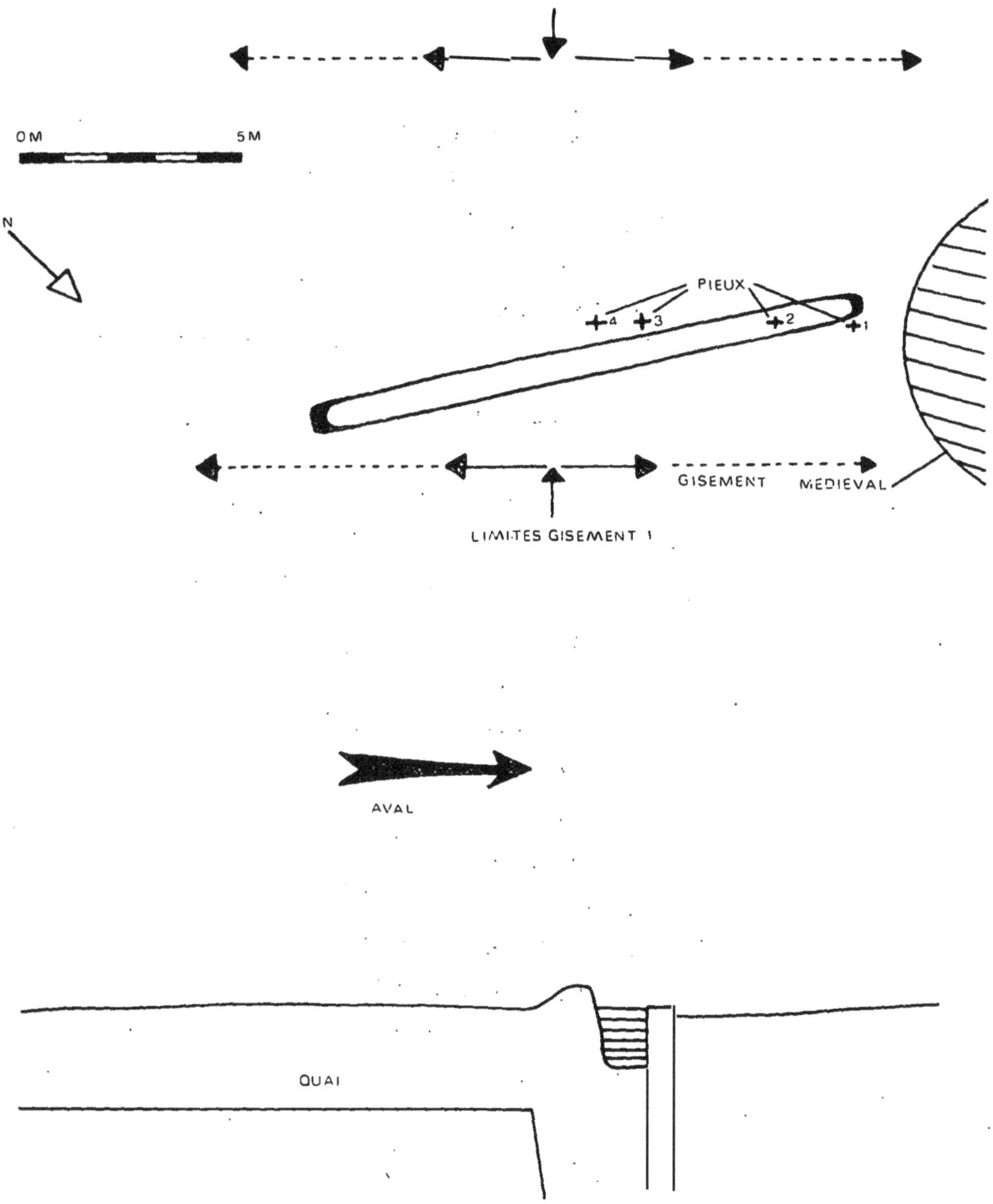

Fig. 7.8 Port-Berteau, plan de la zone I: le port, l'épave de la pirogue, le gisement médiéval

Le port (Fig. 8)

Actuellement, le port constitue la limite sud du hameau de Port-Berteau, qui comprend une dizaine de batiments dont une petite ferme, avec ses dépendances, souvenirs de l'activité agricole du lieu.

Le port se présente essentiellement sous l'aspect d'un quai à deux niveaux. Le plus bas, au bord de la Charente, a une largeur moyenne de 2,50 m. et une longueur de 50 m. Ce quai apparait actuellement comme un simple aménagement de la rive. Cependant, de nombreux blocs de pierre, brisés en plusieurs fragments, sont amoncellés au bas de la pente, dans la rivière. Ils représentent, sans aucun doute, les vestiges d'un soubassement destiné à protéger le quai de l'érosion de l'eau. A l'extrêmité aval de ce quai, un escalier a été taillé dans le rocher. Le second niveau forme une plate-forme de 50 m. de long sur 15 m. de large. La séparation entre les deux niveaux est assurée soit par le sous-sol rocheux lui-même, soit par de gros blocs de pierre, sommairement découpés.

En réalité, la topographie actuelle du port ne représente qu'une image partielle du site. Un examen attentif du plan cadastral, corrélativement à une lecture des archives, montrent que le port était beaucoup plus important et s'étendait sur une cinquantaine de mètres supplémentaires vers l'amont, tout au moins à la fin du XVIIIè siècle.

L'évolution du port, ses limites, son organisation par rapport à la rivière ont été précisés par l'analyse de la répartition de la céramique sur le fond de La Charente.

Le port et la céramique

La céramique, image directe du trafic portuaire, se trouve localisée en trois zones différentes. Le gisement le plus dense (zone 1) s'étend devant le quai moderne. Il se présente sous la forme d'une bande d'une dizaine de mètres de large sur 50 m. de long. Cette bande est orientée dans le sens du courant, à une quinzaine de mètres de la rive droite, c'est à dire au bas de la pente qui relie la rive au lit majeur de La Charente. Si, proportionnellement en surface, la majorité des céramiques moderne est déposée à l'intérieur de cette zone I, des pièces d'époque moderne sont également situées le long de la pente de la rive droite, attestant le lien chronologique entre cette pente et la rive actuelle, limite d'une partie du port moderne. La céramique médiévale ne recouvre pas la même surface. Plusieurs sondages ont révélé une stratigraphie. Celle-ci montre que si la céramique médiévale est présente massivement à l'extrêmité aval du dépôt moderne, elle est beaucoup moins représentée dans la partie amont de ce même dépôt, ce qui tendrait à indiquer une expansion vers l'amont du port moderne (indication confirmée par l'analyse des documents écrits et cadastraux).

A ces données viennent s'ajouter d'autres éléments et en particulier la découverte de quatre pieux. Ceux-ci sont alignés sur un même axe, sensiblement parallèle à la rive droite actuelle. Ils sont à une quinzaine de mètres de la berge, à l'extrémité du dépôt de céramique considéré comme le secteur médiéval du site portuaire. Ils ont une section de 11 cm. de diamètre. Ces pieux sont enfoncés verticalement dans le sol. A ces pieux se greffent de nombreux morceaux de planches et de clous, principalement localisés dans la couche médiévale. Il est raisonnable, dans ces conditions, de relier chronologiquement ces vestiges en bois à l'occupation médiévale du port, correspondant à un ancien tracé de la Charente.

On peut, en résumé, caractériser l'évolution du port de la manière suivante :
- un premier état médiéval où le port comprend des appontements en bois.
- une période d'abandon ou tout au moins de ralentissement de l'activité.
- une troisième étape, moderne, où la rive droite ayant reculée par suite de l'érosion de la berge, le port s'est installé sur la rive droite actuelle, rocheuse et stable. Cette étape correspond, en outre, à une extansion de la surface portuaire vers l'amont.

L'image de ce port, aux installations modestes n'est en rien exceptionnelle. Elle correspond à ce que d'autres fouilles et l'ethnographie ont montré, à savoir, une adaptation minimum de la berge en fonction d'une utilisation maximum de ses capacités naturelles d'aménagements.

Il reste, avant de passer à l'examen de l'épave de la pirogue, un mot à dire sur l'origine de la céramique au fond de la Charente.

Quatre types d'explications peuvent être proposés :
- rejet à la rivière des pièces mal cuites, abîmées pendant le transport de l'atelier à Port-Berteau.
- chutes à l'embarquement.
- naufrages.
- péage marqué par un jet symbolique d'un vase dans la Charente.

<u>La pirogue monoxyle.</u> (fig. 9)

L'épave est située à une quinzaine de mètres du quai moderne de Port-Berteau, le long de la rive droite,

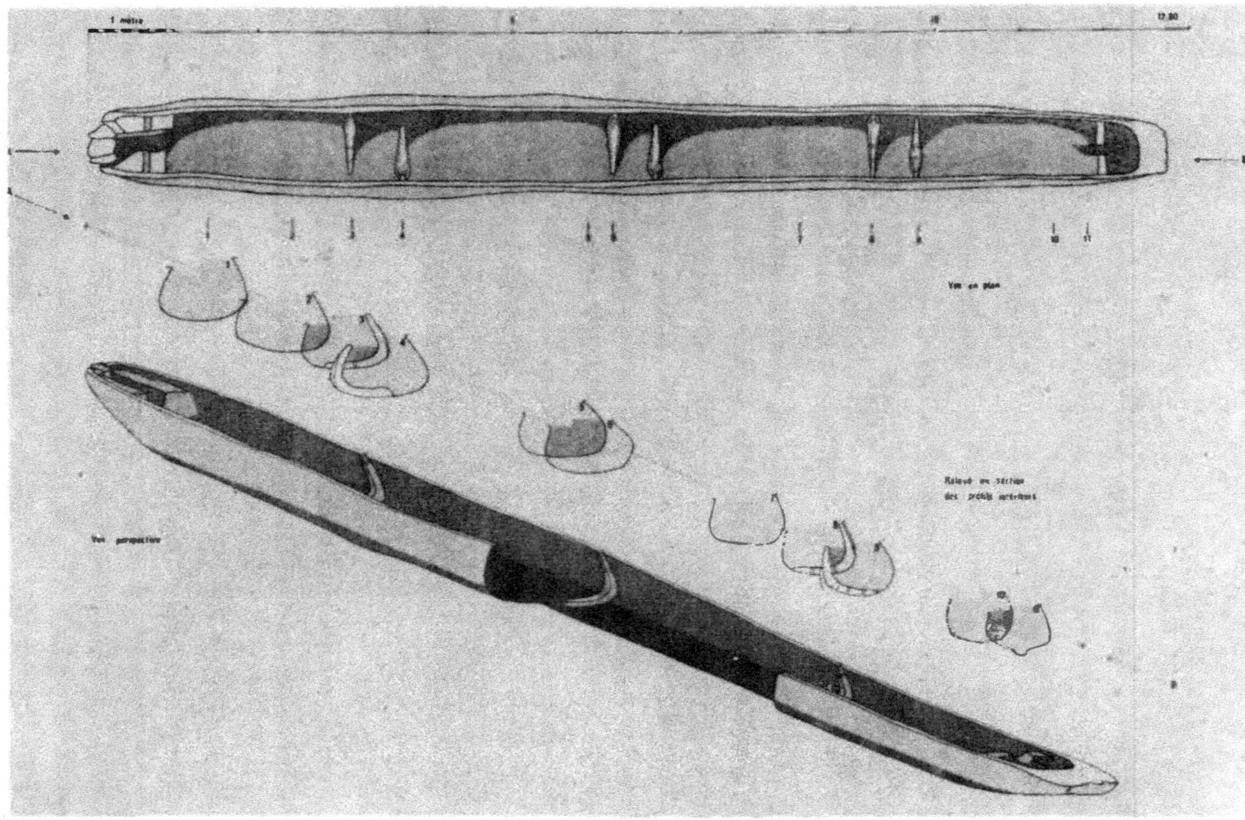

Fig. 7.9 La pirogue de Port-Berteau, XIII-XIVème siècle (Potiers de Saintonge, 1975:111)

dans la partie aval de la zone I. Elle est orientée S.E.-N.W. Elle repose à plat sur le fond de la Charente, à une profondeur de sept mètres.

Cette embarcation monoxyle était, au moment de sa découverte, en bon état de conservation. Précisons que toutes les observations, mesures, relevés ont été effectués dans une eau opaque, avec une visibilité de quelques centimètres, excluant toute possibilité de faire des photographies. Ajoutons que l'étude a été totalement conduite sous l'eau, l'épave n'ayant pas été renflouée.

Les dimensions principales sont les suivantes :

- longueur hors tout : 12.80 M.
- longueur au fond : 11.55 M.
- largeur maximum : 0.75 M.
- creux maximum : 0.70 M.
- rapport l/L : 1/17
- rapport c/l : 1/1,08

En plan, l'embarcation a la forme d'un long trapèze dont l'extrémité amont est plus large que celle d'aval. Il s'agit d'une caractéristique classique de la construction monoxyle maintes fois observées et correspondant à une utilisation maximum de la forme et du volume initial du tronc d'arbre.

En profil, l'extrémité amont est plus élevée que celle d'aval, conséquence logique des caractéristiques précédentes. Les deux extrémités sont taillées en simple biseau.

En section transversale, le fond est en arc de cercle convexe, de même que les flancs.

Revenons à présent en détail sur quelques unes de ces différentes caractéristiques.

<u>L'extrémité amont.</u> (Fig. 10)

Cette partie de la coque monoxyle était endommagée ; un important morceau s'était détaché de l'extrémité ; le

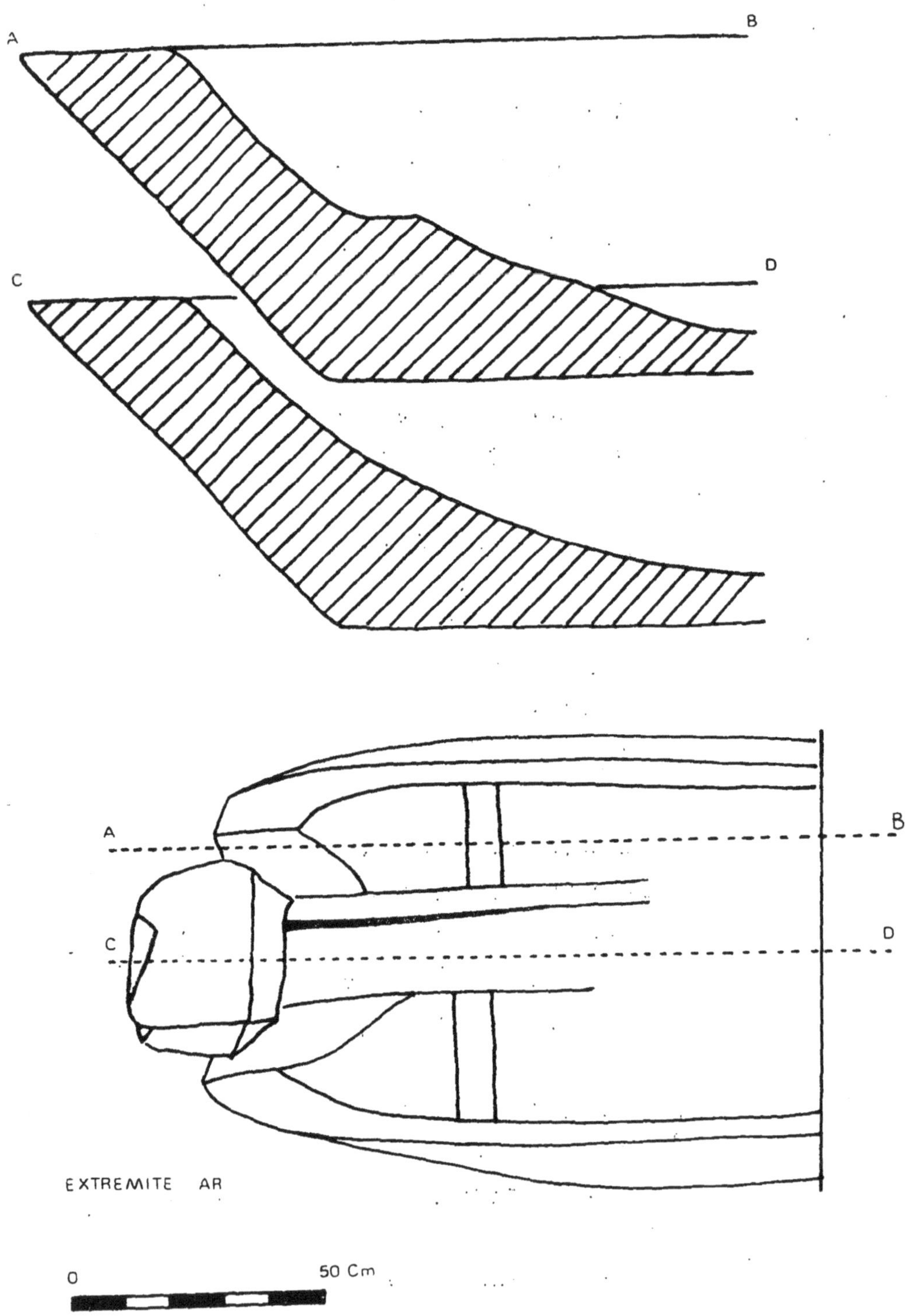

Fig. 7.10 Extrémité arrière de la pirogue

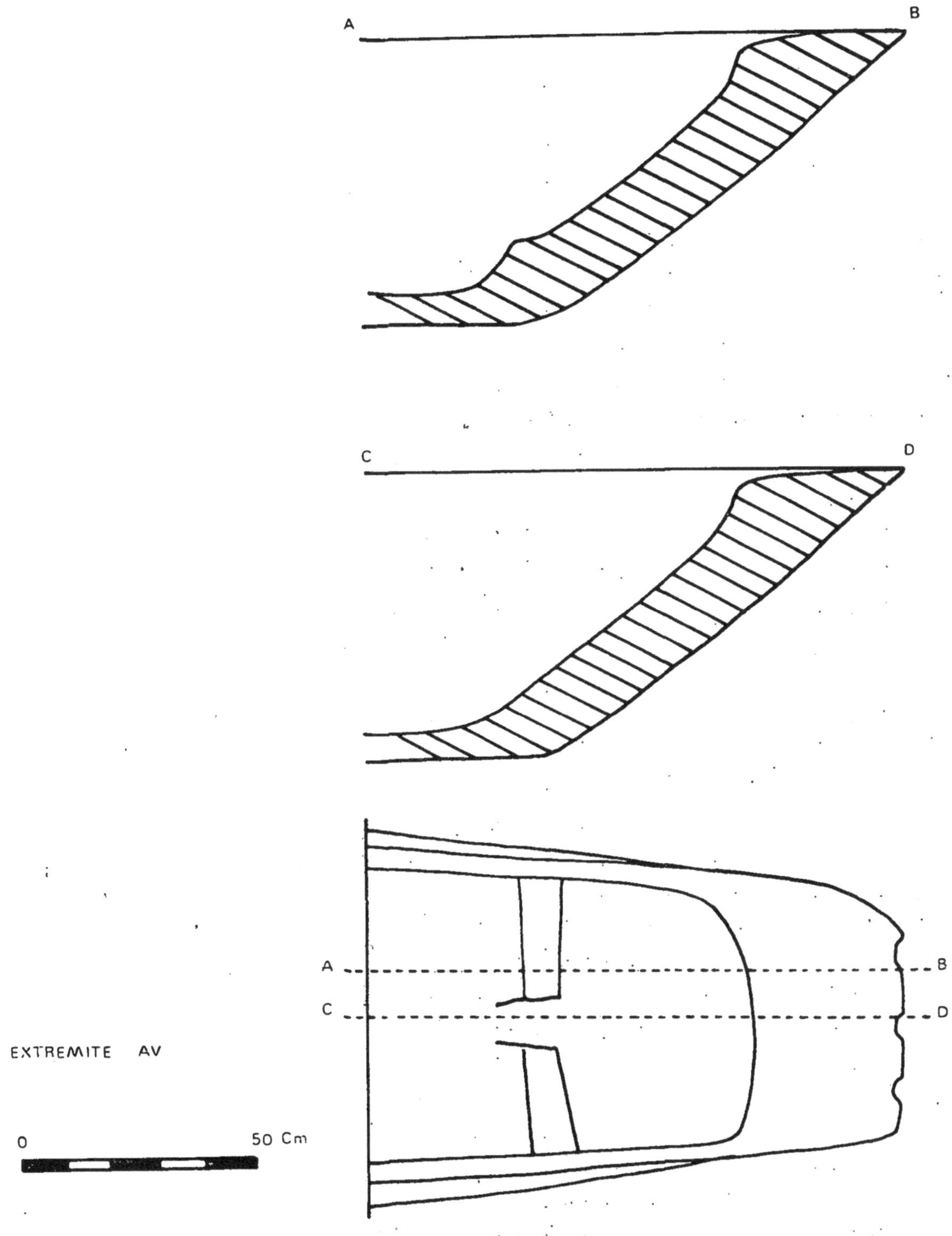

Fig. 7.11 Extrémité avant de la pirogue

fond était fissuré sur soixante quinze centimètres de long. Ce phénomène de dégradation se retrouve fréquemment sur les parties proches des racines du tronc. De plus, les parois étant taillées dans un bois dont le fil est coupé, les fibres deviennent rapidement poreuses et perdent leur résistance. En dépit de cet état de la structure, on peut constater une légère mise en forme puisque la largeur intérieure au plat bord est de 65 CM. à 45 CM. de l'extrémité et de 70 CM. à 1 M. de cette même extrémité. L'épaisseur du bordé est de 6 CM. au plat bord. Enfin, l'intérieur de l'extrémité présente un aménagement sommaire selon le profil interne de la coque.

L'extrémité aval (fig. 11)

Cette partie de l'embarcation était bien conservée. Elle possède, en plan, la forme d'un trapèze de 50x60 CM. de base sur 35 CM. de haut. L'épaisseur du bordé est de 5 CM. au plat bord. L'intérieur présente un aménagement selon le profil en biseau de l'extrémité.

L'examen de ces deux extrémités pose, évidemment, la question de leur situation par rapport à l'avant et l'arrière de l'embarcation. Aucun indice morphologique objectif ne permet, à vrai dire, de répondre sans équivoque à une telle interrogation. Il faut faire un choix. Nous considérerons, à priori, l'extrémité amont comme l'arrière et celle d'aval comme l'avant. En tout état de cause, il nous semble que dans le cas d'une embarcation du type de celle de Port Berteau, une erreur de localisation de la proue et de la poupe n'a pas une grande signification du point de vue de la structure de la coque monoxyle et de sa fonction nautique. Outre une similitude de l'aménagement interne des deux extrémités, l'évidente amphidromie d'une coque au rapport l/L de 1/17 ne peut que confirmer la relative importance de cette orientation : proue - poupe.

Charpente intérieure

La coque monoxyle est étayée intérieurement par une série de trois demi-couples, rapportés et chevillés à la coque.

La position de quatre d'entre eux est précisément connue ; les deux autres, situées au milieu de la coque, n'étaient plus fixés au bordé. Les trous de chevilles dans le bordé n'ont pas été localisés.

Le premier demi-couple M1 est à 3,12 M de l'extrémité arrière, bordé tribord. Le second demi-couple M2 est à 3,72 M de l'extrémité arrière, bordé babord. Le cinquième demi-couple M5 est à 9,27 M de l'extrémité arrière, bordé tribord. Le sixième demi-couple M6 est à 9,82 M de l'extrémité arrière, bordé babord.

Les écarts connus M1-M2 et M5-M6 sont proches : 60 et 55 CM. En outre, la position de l'ensemble M1 et M2, et M5 et M6, par rapport aux extrémités avant et arrière, est sensiblement identique : M1 est à 3,12 M de l'extrémité arrière, M6 à 2,98 M de l'extrémité avant. On ne connait malheureusement pas l'emplacement précis de M3 et M4, ce qui aurait pu permettre de détecter un rythme général au niveau de la charpente, indication précieuse pour comprendre les intentions des constructeurs dans la mise en place des renforts internes.

Chaque demi-couple est fixé par trois chevilles : deux dans le fond et une dans le haut du bordé près du plat-bord. Un examen détaillé des fibres du bois montre que chaque demi-couple est taillé dans une branche fourchue, selon un procédé classique dans la construction monoxyle. La partie en contact avec la coque est la seule a avoir subi un traitement destiné à assurer un bon ajustement du demi-couple à la paroi et au fond. Le dessus est par contre grossièrement équarri. Il faut noter, enfin, que les demi-couples présentent de grandes différences au niveau du travail du bois. Si M1 et M2 sont relativement évolués, M5 et M16 le sont beaucoup moins.

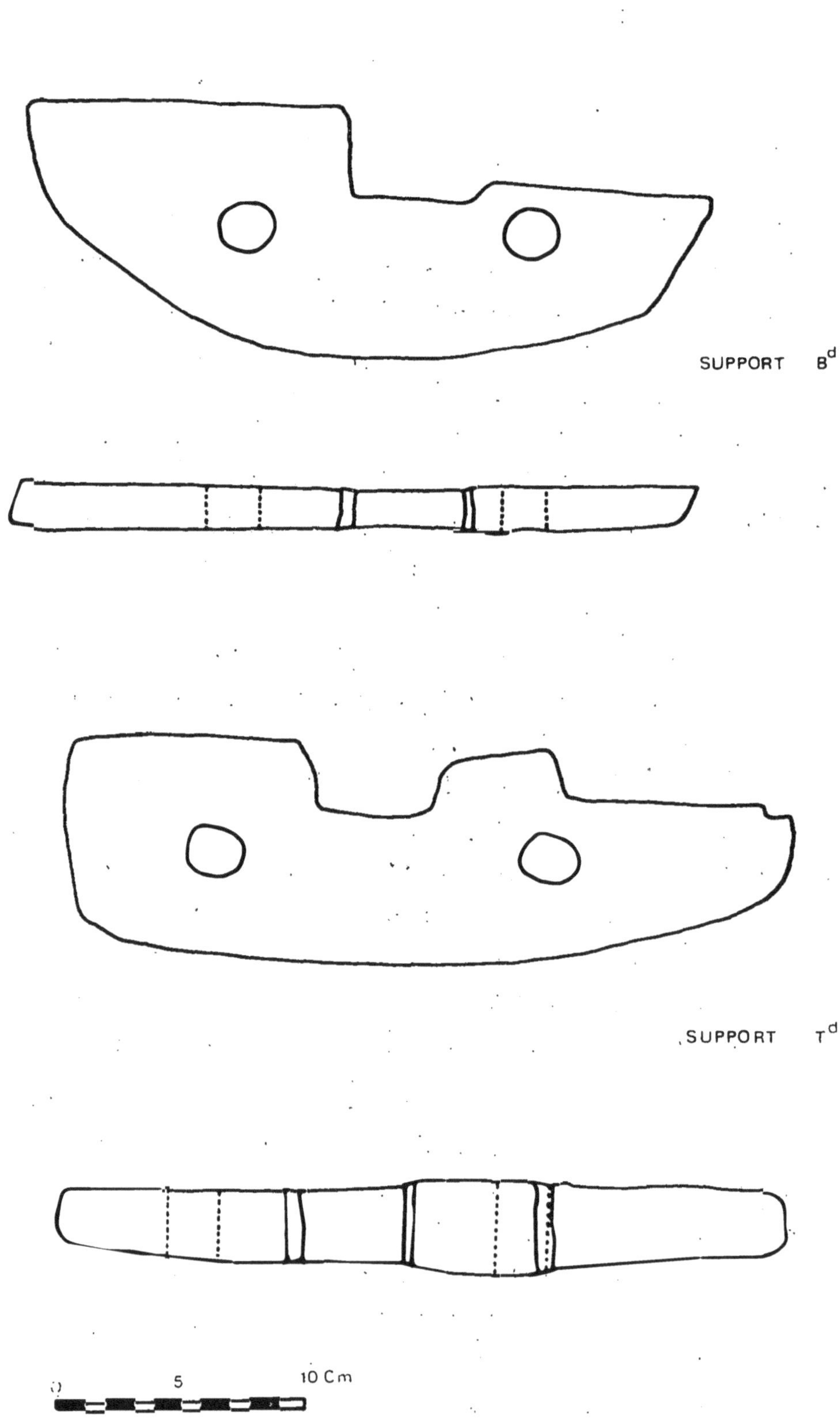

Fig. 7.12 Eléments d'un support transversal

Cette charpente intérieure pose plusieurs problèmes.
-Ces couples ont-ils été insérés à l'intérieur de la coque dès le début de l'utilisation de la pirogue monoxyle ou après un certain temps pour renforcer une coque déjà fatiguée ?
- En outre, tous les demi-couples ont-ils été mis en place en même temps ou existe-t-il un décalage chronologique entre eux ?

Cette charpente ne constitue pas les seuls éléments architecturaux rapportés à la coque monoxyle.

<u>Supports d'un élément transversal</u> (fig. 12)

Au cours de la fouille de la pirogue, deux éléments rapportés ont été trouvés dans le fond de l'embarcation. Malheureusement, ils n'étaient plus fixés au bordé et l'examen de celui-ci n'a pas permis de retrouver les points précis de fixation. Ces éléments se trouvaient à une distance comprise entre 92 CM. et 1,42 M de l'extrémité arrière, l'un contre le flanc tribord, l'autre contre le flanc babord.

Les deux éléments, de forme sensiblement différente, procède cependant d'une même conception. Ils sont constitués d'un morceau de bois, sommairement taillé, de 28 CM de long pour l'un et de 30 CM pour l'autre, d'une hauteur de 10 CM pour l'un et de II CM pour l'autre et d'une épaisseur de 2 CM pour l'un et de 4 CM pour l'autre(dimensions maximums). Dans le milieu de chaque pièce, une encoche de 4 CM x 4 CM est aménagée. Deux trous circulaires de 2 CM de diamètre, séparés par un écart de 10 CM environ, sont percés de part et d'autre de l'encoche rectangulaire. Ils sont destinés aux chevilles de fixation.

L'examen de ces deux éléments pose la question de leur fonction à laquelle il est bien difficile de répondre. Une explication peut être raisonnablement avancée toutefois.

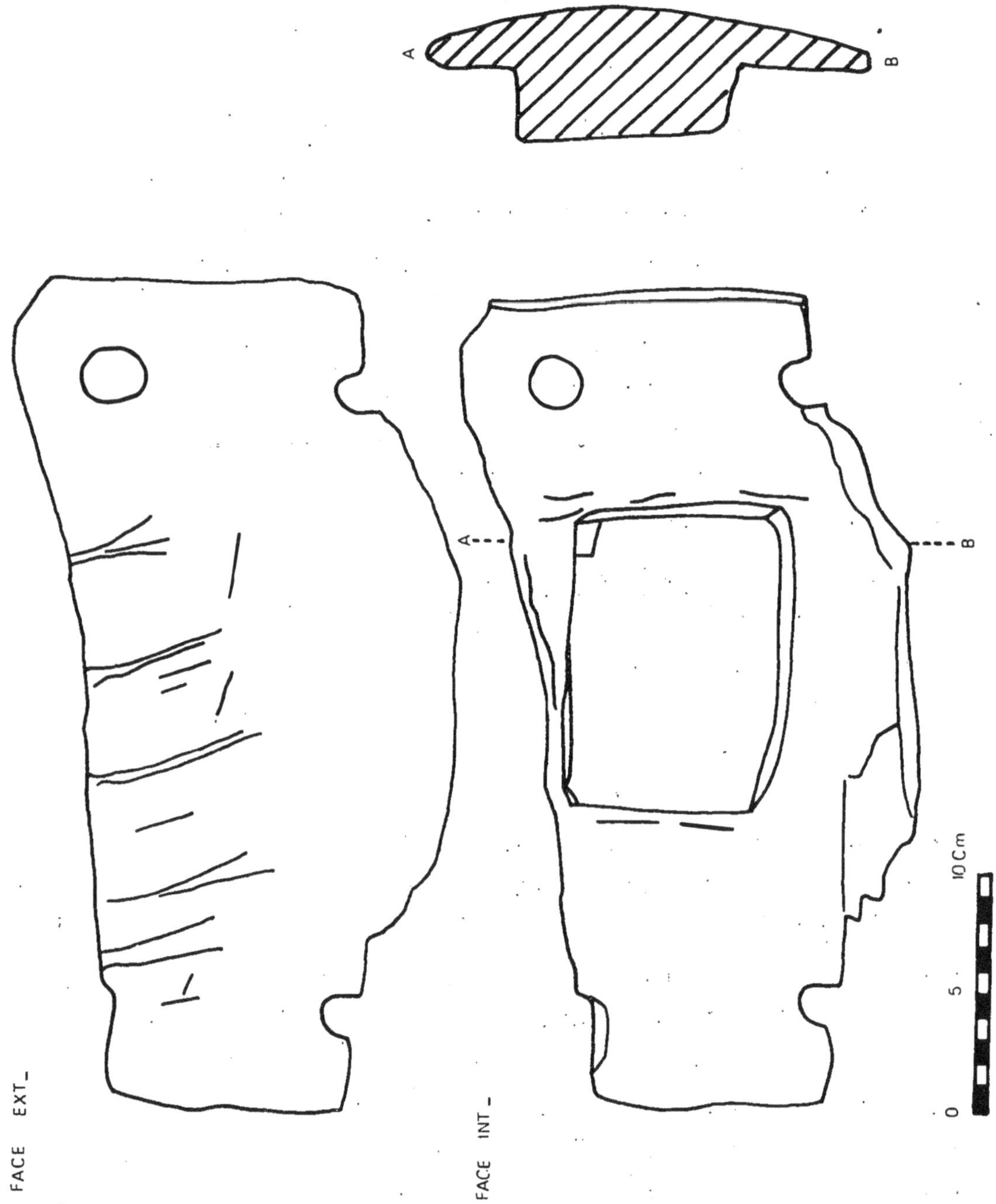

Fig. 7.13 Pièce rapportée au bordé monoxyle

Ces deux éléments transversaux étaient sans doute fixés dans le haut des flancs et servaient de support à un renfort transversal. Deux arguments corroborent cette interprétation : d'une part, ces deux pièces se trouvent au niveau le plus large de la coque monoxyle, à proximité de la base du tronce d'arbre d'origine, partie de la coque qui subit les contraintes les plus importantes ; d'autre part, la présence d'un élément rapporté au bordé près de ces supports peut aussi expliquer ces deux pièces.

<u>Elément rapporté au bordé</u> (Fig. 13)

Cet élément était en place, encastré et fixé au bordé babord, à 3,15 m. de l'extrémité arrière, en face du demi-couple tribord MI. Il se présente sour la forme d'un rectangle très irrégulier,de 35 cm. de long, 20 cm. de large et 2 cm. d'épaisseur moyenne. Les tranches supérieures et inférieures sont très abîmées ; par contre, les tranches latérales sont bien conservées. Les coupes de ces dernières très régulières supposent l'emploi d'un outil à tranchant éffilé. La face externe de cette pièce est courbe tandis que sa face interne est plate. Au milieu de celle-ci, une protubérance rectangulaire de 3 cm. d'épaisseur, 12 cm. de long et 8 cm. de large a été laissé en ressaut au moment de la taille de la pièce. C'est cet élément en relief qui était encastré dans le bordé, fixé par 3 ou 4 chevilles de 2 cm. de diamètre. Remarquons enfin, sur la face externe, une série de 5 à 6 stries verticales de 2 à 6 cm. de long, 1 cm. de large et quelques millimètres de profondeur.

Cette pièce appelle plusieurs questions :

- Quelle est sa fonction? réparation du bordé ou renfort de celui-ci à un endroit subissant des contraintes particulières.

- quelle est l'origine des stries ? lignes de pêche ou filets?
- existe-t-il un lien chronologique entre cette pièce, les supports latéraux et les demi-couples?

Avant de conclure, examinons le problème de la datation de cette embarcation.

Datation de la pirogue de Port-Berteau

Aucune datation par des moyens physiques, dendrochronologie ou Carbone 14, n'a été utilisée. La datation a été précisée uniquement par l'analyse du contexte archéologique général. En effet, la structure de l'embarcation ne peut être considérée comme un indice chronologique évident. Il ne fait aucun doute que l'embarcation de Port-Berteau est antérieure au XVIIème : une couche sépare nettement la strate moderne de celle où se trouve la pirogue ; plusieurs indices (céramique en particulier) laissent à penser que la pirogue appartient à la fin du XIIIè - début XIVè siècle. Il faut noter,enfin,que la présence d'un pieu traversant le fond de l'embarcation implique que cette dernière est antérieure au pieu et au quai en bois auquel il appartient.

Conclusion

La pirogue médiévale de Port-Berteau présente, au niveau de la coque, les caractéristiques classiques de la construction monoxyle avec sa forme trapézoïdale, ses extrémités en biseau. Sa charpente antérieure de demi-couples, ses deux supports d'un renfort transversal, son élément rapportéau bordé, tous chevillés, dénotent une connaissance manifeste d'une technologie de l'assemblage fondamentalement distincte de celle du pur monoxyle. Le percement très régulier de trous de cheville dans l'élément rapporté au bordé implique en outre l'emploi d'un outillage élaboré, du type tarière.

Il est frappant de remarquer le contraste entre cette embarcation monoxyle, antérieure de quelques décennies à la phase d'exploitation du port médiéval et la qualité technologique de la céramique exportée et d'autre part le circuit d'échanges élaborés conduisant la céramique de Port-Berteau en Angleterre.

Notons, d'ailleurs, que cette construction monoxyle se prolonge jusqu'au XVIIè siècle : un texte de 1644 (Sépulchre, 1978 : 25) précise qu'un affluent de La Charente, la Touvre, "... ne peut porter un bateau fait de diverses pièces... il faut nécessairement qu'il soit d'une seule pièce de bois." Or, le XVIIè siècle est la période où La Charente voit se créer une batellerie attestée, entre autre, par la gabarre représentée dans l'Album de Jouve (1679) (Jouve, 1679 : pl.15). Ce phénomène de juxtaposition de traditions de constructions navales très différentes, les unes archaïques, les autres élaborées, n'est pas unique. On le retrouve, par exemple, sur l'Adour jusqu'au XIXè siècle (Beaudouin, 1970).

Le port lui-même présente des aménagements très sommaires, sans comparaisons aucune avec le haut niveau d'échanges dont il est le lieu.

Cet exemple de Port-Berteau constitue une illustration très claire d'un ensemble (port et embarcation) qui se rattache à une tradition essentiellement rurale. En dépit de son lien avec un axe économique dominant essentiellement maritime, Port-Berteau reflète l'image d'un port pronfondément intégré au paysage rural qui l'environne.

Eric RIETH

REFERENCES

Beaudouin, F., 1970. <u>Les bateaux de l'Adour</u>, Bayonne

Chapelot, O. et J., 1973. <u>Le cimetière des indiens Tonica (Louisiane) ; étude d'un matériel archéologique d'origine européenne</u>, rapport dactylographié (inédit)

Craeybeckx, J., 1958. <u>Un grand commerce d'importation : les vins de France aux anciens Pays-Bas</u>, Paris

Dunning, G.C., "The trade in Medieval Pottery around the North sea". <u>Rotterdam Papers</u>, Rotterdam : 35-38

Dunton, J.V.N., 1971. "Céramiques françaises du XVIIIème siècle retrouvées en Nouvelle France". <u>Cahiers de la céramique, du verre et des Arts du Feu</u>, 48 : 12-25

Jouve, J. (attribué à), 1679. <u>Desseins des différentes manières de vaisseaux que l'on voit dans les havres, ports et rivières depuis Nantes jusqu'à Bayonne qui servent au commerce des sujets de sa Majesté</u>. manuscrit, Service historique de la Marine, Paris

Orsaud, D., 1975. <u>Le trafic d'un port fluvial sur la Charente, Port Berteau. Etude de la céramique des XIIIè-XVIIIè siècles découverte en fouille subaquatique dans le lit de la rivière</u>. Maîtrise de l'Université de Paris I

<u>Potiers de Saintonge, huit siècles d'artisanat rural</u>. Catalogue de l'exposition, Musée des Arts et TRaditions populaires, Paris, 1975

Rieth, E., 1973. <u>La Charente. Essai d'étude de sa navigation. Apport de l'archéologie subaquatique à partir de l'exemple du site du port fluvial de Port-Berteau</u>. Maîtrise de l'Université de Paris I

Sepulchre, B., 1978. *A la recherche des gabariers de la Charente.* Jarnac

7. PIROGUE MONOXYLE ET PORT FLUVIAL DE PORT-BERTEAU

Eric Rieth

English Summary

Port Berteau is a small harbour on the River Charente, between Saintes and Rochefort. The site was excavated by underwater methods during the three years 1971 to 1973. This excavation was part of a vast programme conducted by the Musée des Arts et Traditions Populaires of Paris, under the scientific direction of Jean Chapelot (Université de Paris I) to study the ceramics produced in Saintonge, during the medieval and post-medieval periods, and their transport by river and sea.

Archaeological finds suggest that Port Berteau was a specialised harbour where pottery was loaded into boats and sent to North Europe during the medieval period and to the French colonies in America in the post-medieval period.

The natural conditions of the settlement were studied; secondly its structural evolution was analysed through the horizontal distribution of the artefacts lying on the river's bed. Archaeologically speaking, little remains of the harbour were discovered: a few set timbers and planking, but no timber revetment. In front of the quay, a great logboat (13th-14th century) was found, with six half-frames fixed by wooden treenails to the planking. This logboat was studied in relation to the river's local trade.

KEYWORDS

Medieval and later.

France; River Charente; Port Berteau.

River harbour; logboat.

Boatbuilding characteristics; trade.

8. MEDIEVAL RIVERFRONT REVETMENT CONSTRUCTION IN LONDON

Gustav Milne

Summary

Recent waterfront excavation has revealed many examples of revetments constructed during riverfront reclamation in the City of London between the 13th and 15th centuries. The structures are classified, described and illustrated with carpentry details. The contemporary level of the Mean High Water Neaps was deduced by examining the height at which the timberwork decayed, and the type of craft which operated from the frontage is suggested.

The City of London was a port of international importance for much of its 2,000 year history. Since the 1950s however, the volume of trade brought to its wharves decreased sharply, and now the City no longer handles riverborne merchandise. It has subsequently developed a new role as an insurance and banking centre, and virtually the entire waterfront has been—or will be—subjected to large scale redevelopment involving the complete destruction of the Victorian warehousing. A detailed examination of this area is obviously important for archaeologists intent on understanding the growth of Britain's principal port,so, since 1972, the Guildhall Museum and its successor the Museum of London have excavated six sites between Thames Street and the river.[1] From Blackfriars in the west to the Tower in the east, the sites (with their codes) are: Mermaid Theatre (THE); Baynards Castle (BC); Trig Lane (TL); Seal House (SH); New Fresh Wharf (NFW); Custom House (CH). These sites are all marked on Fig. 5.1 with the exception of THE, which is to the west of Baynards Castle, and SH, just west of London Bridge.

One of the most important results of this intensive work was the demonstration that the majority of the land south of Thames Street had been reclaimed at the expense of the river between the 12th and the 16th centuries. Reclamation was achieved by erecting a timber revetment upon the foreshore to the south of the existing frontage and infilling the intervening area with dumps of domestic refuse, sealed by a stone or gravel surface. It was possible to study the mechanics and chronology of encroachment on three adjacent properties on the Trig Lane site, where it was shown that reclamation was the prerogative of each individual property owner or occupier, and was not a corporate concern. This situation produced a structurally varied and much indented frontage, in which almost every property was served by its own private stair to the foreshore.

The revetments themselves often survived to heights of 2 m or more. Over twenty have been recorded in London and although all were different, they clearly belonged to the "vertical tradition" of revetment construction, and not to the "horizontal tradition", well-known from the Bryggen excavations in Bergen (Norway) for example. Vertical revetments in general may be divided into three classes, UNBRACED; FRONT-BRACED and BACK-BRACED. Of these, the London examples described below all fall into the latter two categories, and date from the 13th to the 15th centuries.

The Custom House site produced a simple type of front-braced revetment incorporating vertical posts driven into the foreshore against which horizontal planking was nailed (Fig. 8.1A). Part of the cladding was in fact a section of a clinker built boat described in the paper by Marsden in this volume. The W III revetment at Seal House was similar in form, but enjoyed the sophistication of a base-plate into which the vertical posts were set (Fig.8.1B). On this example, the planking was fixed to the posts with wooden pegs instead of iron nails, and on an earlier revetment on the same site (W II), the ends of the horizontal planking were set into grooves cut into the edges of the vertical posts, a feature also noted at Baynards Castle and New Fresh Wharf. The later 14th century structure at Custom House was a more elaborate revetment with its principal and subsidiary base-plates tied by additional plates, into which the feet of the shores (or braces) were chase-tenoned (Fig.8.1D).

On the Mermaid Theatre site evidence of a revetment incorporating both front and back-braces was recovered (Fig.8.2A), similar to the G3 structure at Trig Lane. The latter had back-braces spaced at 3 m intervals as well as more closely spaced front braces, as can be seen on Fig.8.3. Thereafter, the later revetments on this site only employed back-braces to support the facing which was composed—in the G7 example—of horizontal planking (Fig. 8.5A), or vertically set stave-work in the G10 and G11 structures (Fig.8.4).

The G10 and G11 structures also exhibited an additional feature not observed in any earlier examples: both were built in two distinct levels, or stories. It would appear that the upper level of timberwork was more prone to decay than the lower, and thus needed to be replaced more often. Because the upper level was structurally distinct from the lower, it could be replaced as a unit without affecting the stability of the foundation. Such a bipartite structure was an improvement on the earlier revetments which had to be replaced completely when the heads of the timbers began to decay. The G12 structure for example replaced the original upper section of the G10 revetment some 60 years after the lower section had been erected (Fig. 8.5C).

Observation of structural timberwork on the presentday Thames foreshore suggests that the zone of decay is probably confined to the area between the Mean High Water Neaps and Mean High Water Springs, in London c. +2.70 and 3.90 m above OD. Above this zone, timberwork is permanently exposed above the fluctuating waterline, while below it, it is submerged twice daily. The decay may be simply the result of differential swelling and shrinking above and below the zone, but the role played by algae and other organisms living in this zone is a subject of current research.

An examination of the riverfront revetments can therefore be used to establish the level of the contemporary Mean High Water Neaps. The London evidence suggests that in the 14th and 15th centuries it was above +1.20 m

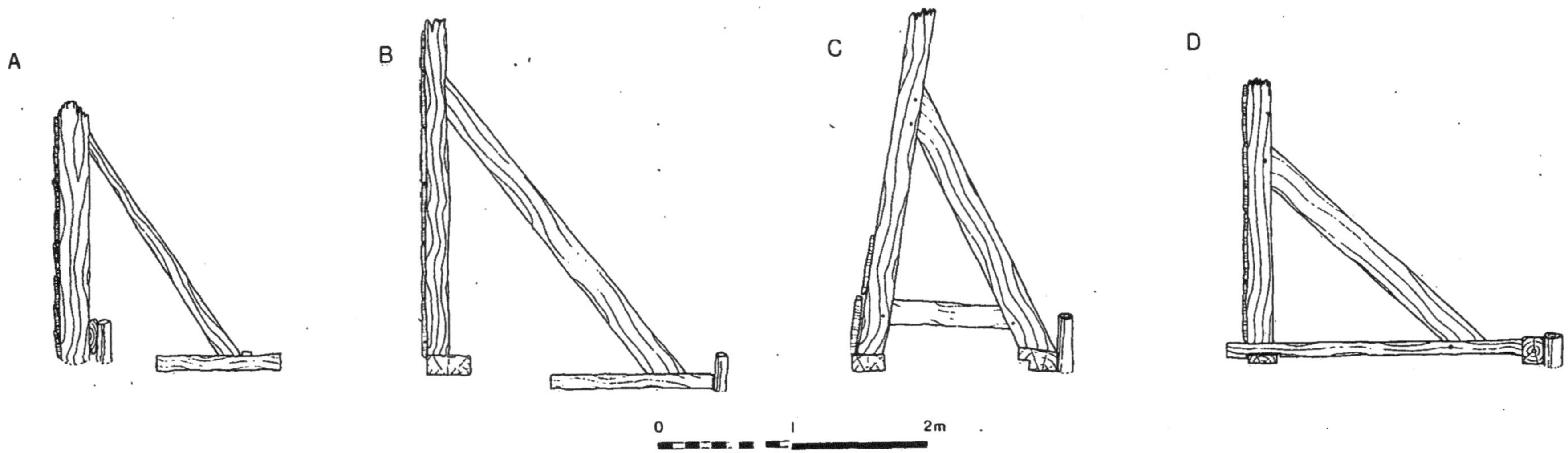

Fig. 8.1 Side elevations of 13th-14th century front-braced revetments from London: river to right. A. Custom House I; B. Seal House WIII; C. Trig Lane G6; D. Custom House II

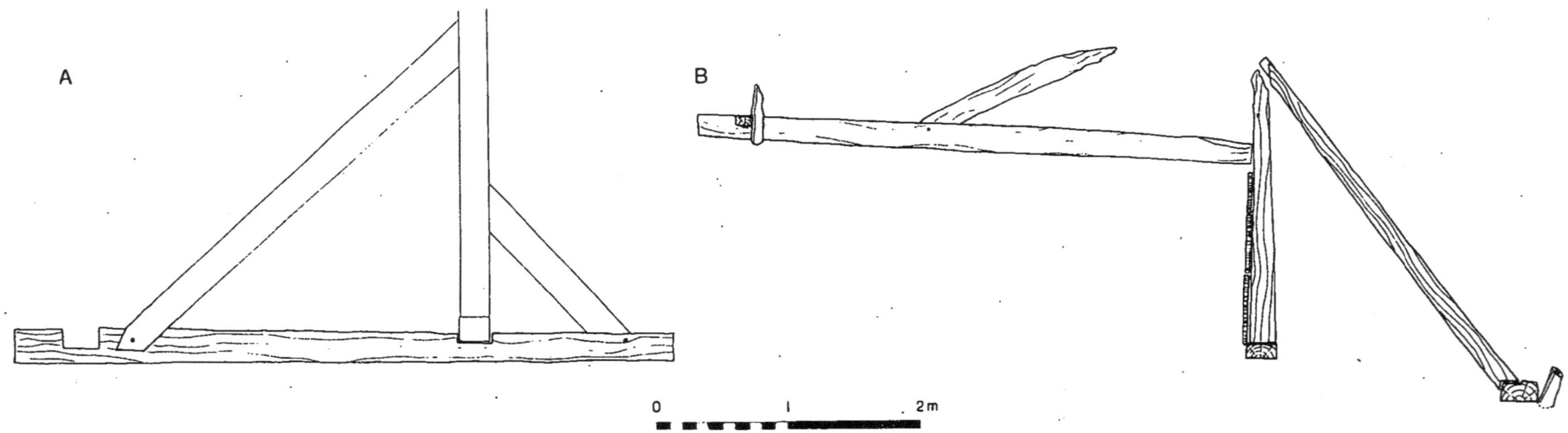

Fig. 8.2 Side elevations of 13th-14th century front-and-back-braced revetments from London. A. Mermaid Theatre; B. Trig Lane G3

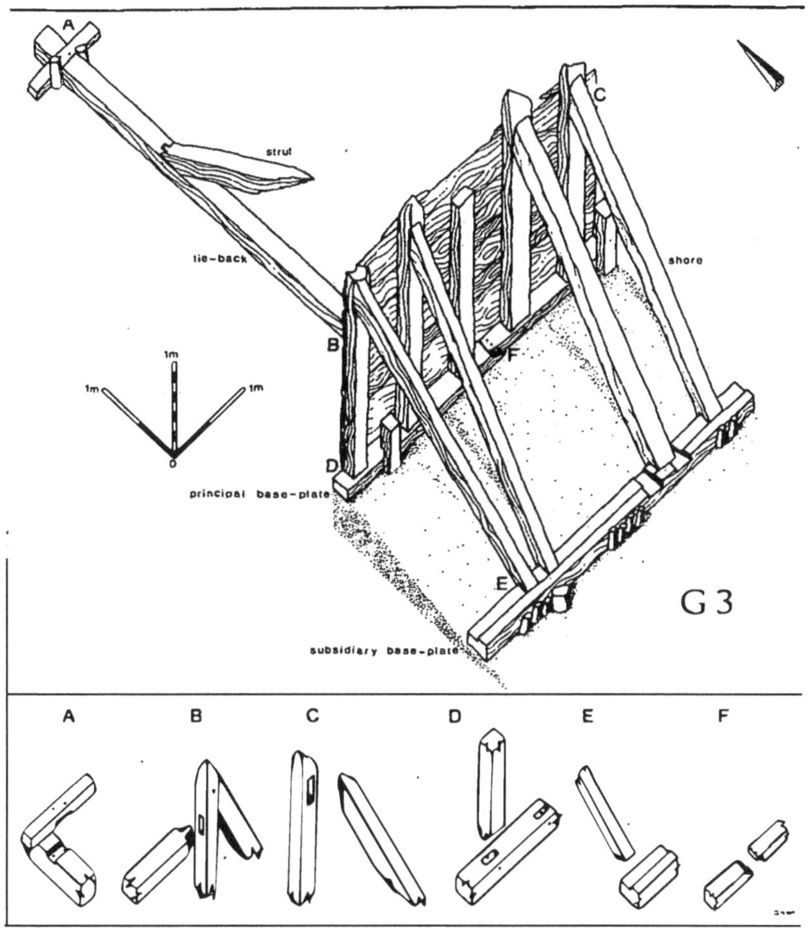

Fig. 8.3 Semi-reconstruction of front-and-back-braced G3 revetment at Trig Lane, London. River to S. Joint details: A. Pegged half-lap; B and D. Central tenon; C. Chase tenon; E. Bird's mouth abutment; F. Edge-halved scarf with square vertical butts and two face pegs

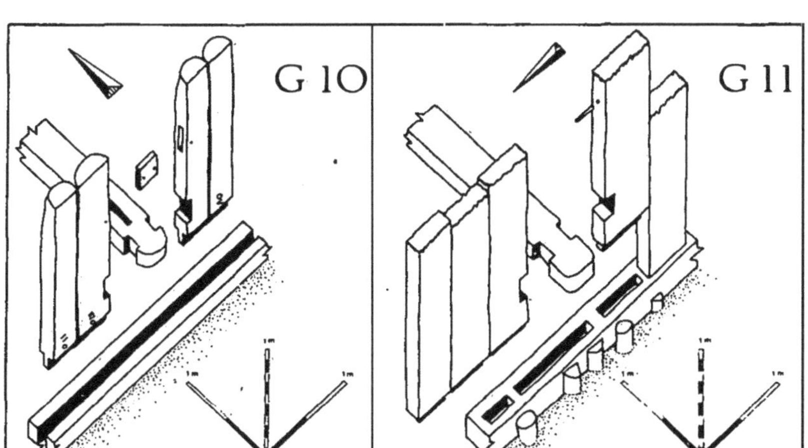

Fig. 8.4 Part of 14th century back-braced revetments from Trig Lane, London. River to S for G10, W for G11

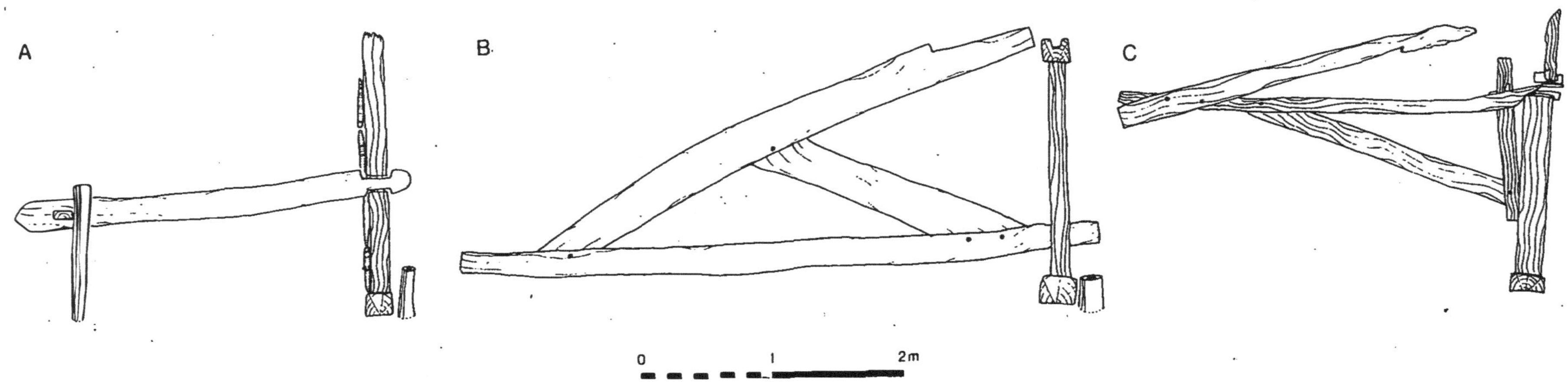

Fig. 8.5 Side elevations of 14th-15th century back-braced revetments from Trig Lane, London. River to right. A. G7; B. G11; C. G12

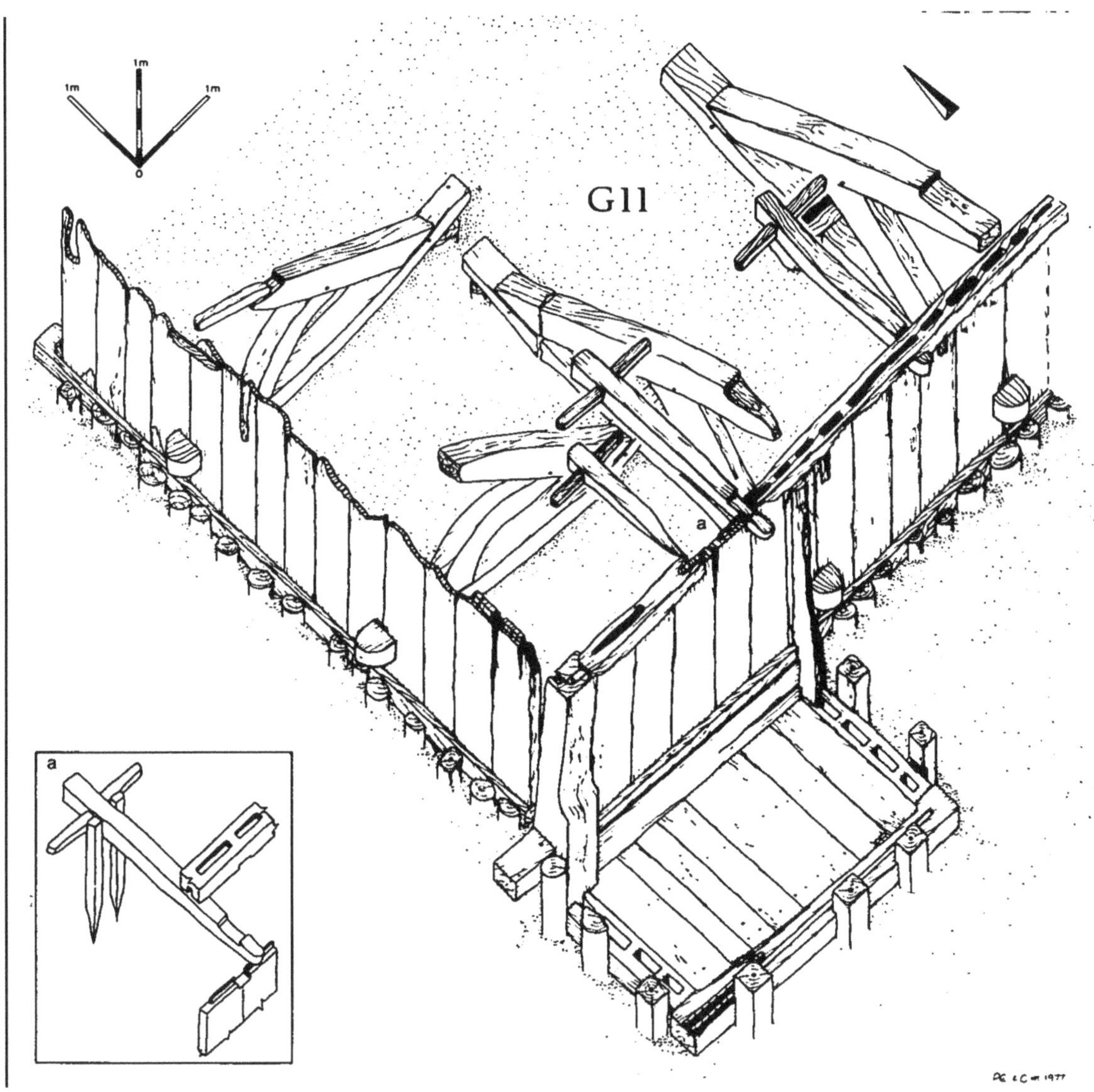

Fig. 8.6 Late 14th century back-braced revetment from Trig Lane, London as found. The upper level of vertical planking had been removed

but below +1.50 m OD. The Mean High Water Springs could not have been above +2 m OD, as occupation surfaces were found at this level.

A detailed record was made of joints and carpentry techniques employed in the construction of many of the revetments. Standard and chase mortises and tenon were the most common method of joining shores to base-plates, posts and tie-backs, although notched lap joints were noted on the example from the Mermaid Theatre site. Other types of lap joints were observed on the G3 (Fig.8.3) and G12 revetments (Fig.8.5C) at Trig Lane. Three 14th century revetments from this site utilised a system of edge-trenching at the junction of the horizontal tie-back (G7) or back-brace base-plate (G10 and G11) with the vertically set posts or staves (Fig.8.4). The unwithdrawable joint of exceptional strength thus produced is quite unknown in any medieval building on dry land, although it was obviously very common on the riverfront: the closest parallel would seem to be the joint between the beam and the ribs on such vessels as the Bremerkogge (see paper by Ellmers in this volume), which was contemporary with the G11 revetment.

Free tenons were sometimes used to join adjacent vertically set stave members on the G10 structure, which also had a series of assembly marks incised upon the face of the timbers, demonstrating that it was a prefabricated structure which had been erected from west to east (Fig. 8.4). Examples of sawn timbers (G11), timbers halved by splitting (G10) and radially split planking were also noted.

There was evidence for diagonally set river stairs descending onto the shelving foreshore to the south of the revetted properties, sometimes incorporating the trestle-based jetty structures known as "bridges". River craft need not therefore have tied up directly alongside the 2 to 3 m high revetments, but at the foot of the stairs. As three contemporary river stairs some 6 to 7 m apart projected from the 14th and 15th century frontage excavated at Trig Lane, it is suggested that only small river vessels could have operated in this area.

Thus the analysis of the medieval riverfront revetments from the City of London has demonstrated that such features are worthy of study in their own right; that it is possible to calculate the level of the contemporary Mean High Water Neaps by isolating the relevant decay zone; and that the general nature of the river craft which may have operated from the waterfront associated with the revetments may be suggested.

Note.

1. Site supervisors were P. Herbert (THE); P. Marsden (BC); G & C. Milne (TL); J. Schofield (SH); J. Schofield & L. Miller (NFW); T. Tatton-Brown (CH).

REFERENCES

Herteig, A., "The excavation of Bryggen, Bergen, Norway" in R. Bruce-Mitford (ed) Recent Archaeological Excavations in Europe (1975) 65-89.

Hobley, B. & Schofield J., (eds). 1977. "Excavations in the City of London: first interim report, 1974-5", Antiq. J., 57.

Milne, G. & Milne, C., 1978, "Excavations on the London Waterfront at Trig Lane, London 1974-6" Med Archaeol, 22:84-104.

Milne, G. & Milne, C., 1979. "The making of the London Waterfront", Current Archaeol, 66:198-204.

Milne, G. & Milne, C., in preparation. "Excavations at Trig Lane City of London 1974-6: Part One", L.A.M.A.S. Special Paper.

Schofield, J., 1975, "Seal House". Current Archaeol, 49:54-57.

Tatton-Brown, T., 1974, "Excavations at the Custom House site, City of London, 1973" L.A.M.A.S. 25:117-219.

Tatton-Brown, T., 1975, "Excavations at the Custom House site, City of London 1973: Part Two", L.A.M.A.S., 26:103-170.

Port of London Authority, Handbook of Tide Tables (1979).

L.A.M.A.S. = London and Middlesex Archaeological Society.

KEYWORDS

Medieval.
Britain; London.
Timber structures; waterfront; revetments.
Classification; joint details; Mean High Water Mark; reclamation.

9. THE ARCHAEOLOGY, HISTORY AND ARCHITECTURE OF THE MEDIEVAL PORTS OF THE EAST COAST OF ENGLAND, WITH SPECIAL REFERENCE TO KING'S LYNN, NORFOLK

Helen Clarke

Summary

The medieval ports on the east coast of England owed their prosperity to overseas trade. Evidence for this can be traced in historical documents, particularly from the fourteenth century and later, from archaeological excavation and from the medieval, commercial, predominantly waterside, buildings which still stand in some ports. King's Lynn, Norfolk, is especially well provided with all three types of evidence, and study of them has illustrated aspects of the development and growth of a medieval east coast port.

During the past two decades a number of medieval ports on the east coast of England have been the subject of archaeological study. The initial impetus was given by work in King's Lynn, Norfolk, aspects of which will be discussed in more detail below, but more recently there have also been excavations in, for example, Hull, Ipswich, Harwich and Sandwich. Waterfront sites are beginning to be seen as important objectives in urban archaeology and their changing use, position and importance is becoming recognized and understood. Archaeological evidence for trade has also received attention recently. In this paper the east coast ports will be set against their historical background and certain aspects of the archaeological investigation of one of them, King's Lynn, will be discussed in detail.

Fig. 9.1 shows the medieval ports on the east coast which are known to have had international trading connections. Evidence for their trade may be from historical documents (mentioned in Customs Accounts, known to have had fairs or markets and so on), or as a result of archaeological investigation (the discovery of imported goods, notably pottery). The map is a record of the first stage of the author's work on east coast ports, and will need emendation and expansion as research progresses, but in the present state of knowledge several observations may be made: most east coast ports are post-conquest foundations and lie closer to the coast than do the pre-conquest ports which are fewer in number and most often lie on rivers.

The ports founded on the east coast after the Norman conquest almost all came into existence during the two hundred years from the end of the eleventh century when both the population and economy of Europe were increasing. Most proved to be successful plantations with close links across the North Sea to Norway, the Baltic, the Low Countries and France and they have flourished

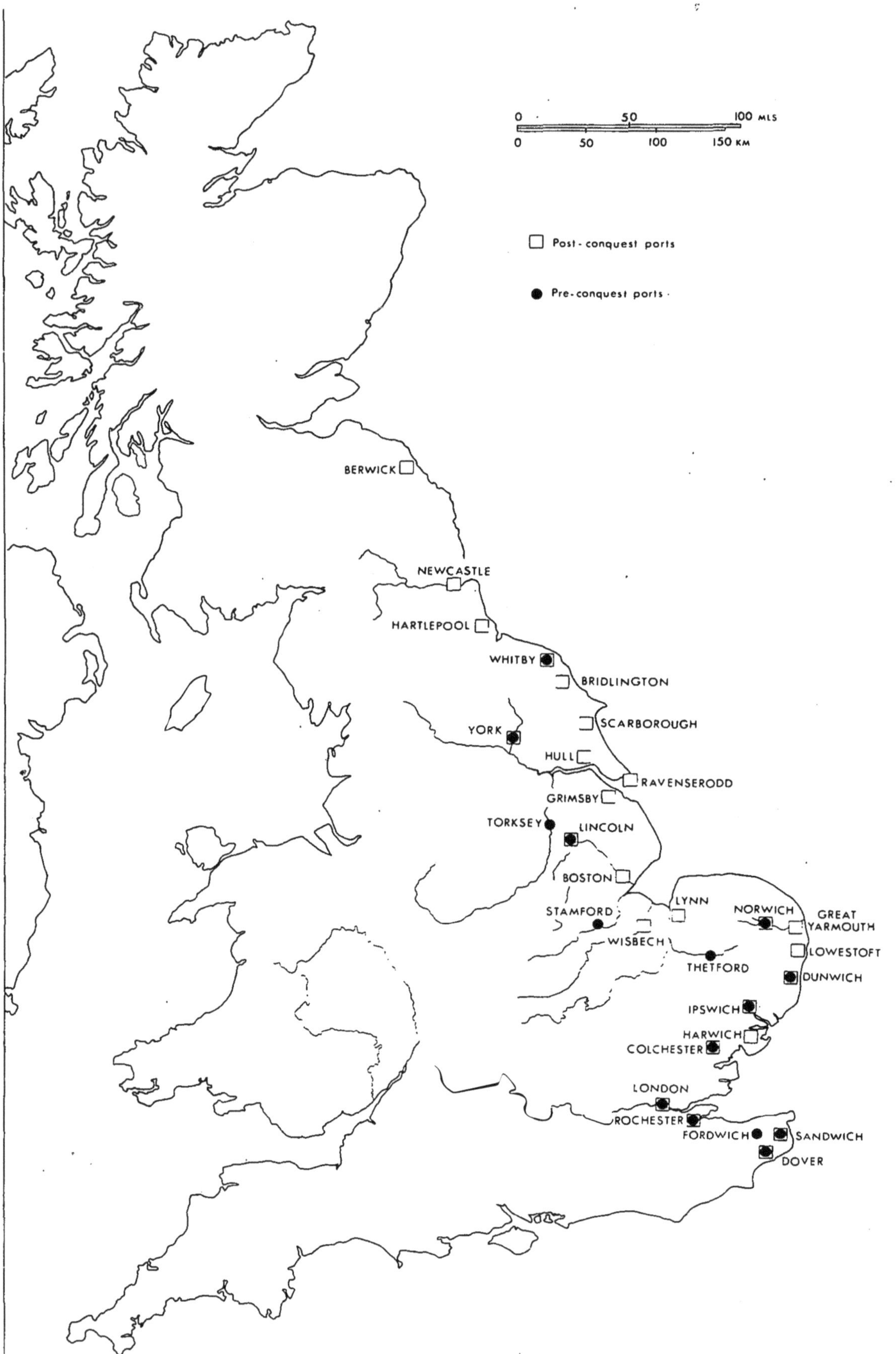

Fig. 9.1 East coast ports with international trading connections in the Middle Ages

until the present day, but some, such as Ravenserodd on Spurn Head and Dunwich in Suffolk (a pre-conquest port which continued into the high middle ages) failed after natural disasters, and no longer exist.

The overseas trade which stimulated the foundation and encouraged the development of the east coast ports is well attested in historical sources although little documentary evidence is preserved for the twelfth and thirteenth centuries, the period when they were beginning to grow and flourish. At this time the fairly scanty sources indicate the export of raw wool to the Low Countries and grain to Norway, largely through the port of King's Lynn (Carus-Wilson 1962-3, 185). Other exports from England to Norway are listed by King Sverri in his speech to the men of Bergen at the end of the twelfth century: wheat, honey, flour, cloth, linen, flax, wax and cauldrons. The speech also records the presence of Germans in Bergen. They are castigated by King Sverri for bringing wine to Norway in exchange for stockfish and butter. The Englishmen, on the other hand, are thanked for the useful wares which they brought (Sephton 1899, 129-30).

Documentary evidence for the international trade of the east coast ports increases after the late thirteenth century with the 'Great and Ancient Custom' on wool instituted in 1275 (Salzman 1931, 354) and the 'New Custom' of 1303 (Carus-Wilson 1962-3, 192) which imposed taxes on all goods imported into England by foreign merchants. From then onwards there are lists of the amount of wool being exported through English ports and the goods which were imported in return. The lists are, of course, incomplete, not least because of the exclusion of the goods imported by English merchants, but they are good indicators of a brisk import trade and suggest that the fourteenth century was the period of greatest prosperity for the medieval ports of the east coast. Stockfish, deal boards and masts, goat skins, furs and litmus were imported from Norway (Darby 1951, 326; Helle 1967; Nedkvitne 1976; Salzman 1931, 364). Similar materials were brought by Baltic traders, with the addition of metal and metal goods, raw materials for cloth manufacture, wax, furs and beer (Salzman 1931, 284). Woollen cloth was imported mainly from Flanders and Brabant (Darby 1951, 317). Wine (James 1971) and building stone (Dunning 1977, 331-47) came from France.

In exchange for all these goods the east coast ports exported mainly perishable materials, particularly wool and wool felts, and their fourteenth-century prosperity was built on this (Power 1941). By the end of the fourteenth century, however, the wool trade became increasingly concentrated in London and the east coast ports consequently declined in wealth and importance. Exports of grain to Norway, coal to the Baltic, the Low Countries and France, and miscellaneous manufactured goods to Iceland continued throughout the fifteenth century (Carus-Wilson 1962-3, 186, 199), but the trade of the North Sea was becoming progressively more dominated by the German Hanse, to the detriment of the ports of the English east coast. Hanseatic Steelyards were founded in King's Lynn, Boston and Hull in the second half of the fifteenth century and helped to keep the Baltic connections of these towns alive (Carus-Wilson 1973; Purcell 1973), but in general the east coasts ports reverted largely to the coastal trade of which we have so little documentary evidence.

The documentary evidence for the existence of the pre-conquest ports, let alone their international trade, is very much slighter than that for the later period and often our only information comes from their having been mints (Ballard 1914, 44-5). Some, such as Dunwich (Beresford & Finberg 1973, 166) Ipswich (Dunmore et al. 1975) and York (Hall 1978, 32) have better documentation, but even so archaeological work in these sites will probably remain our best guide to their trading activities. Other settlements such as Stamford (Mahany et al. forthcoming), Thetford (Davison 1967; Dunmore & Carr 1976) and Torksey (Barley 1964) lie so far inland that their attributions as ports could be doubted, but all show evidence for international contacts and continental examples show that there was no disadvantage (and possibly a positive advantage) in international entrepôts lying far away from the sea. Dorestad in the Netherlands is a particularly good example of this (van Es 1969).

It was mentioned on p. 155 that post-conquest ports seem to lie on the coast whereas many pre-conquest ports lie on rivers some distance from the sea. This is a phenomenon of continental as well as English ports (Clarke in press) and the reasons for it may be complex. An increase of trade in bulky goods in the eleventh century may have led to (or gone hand in hand with) the development of large cargo vessels such as that typified by the Bremen cog (p. 1). Such vessels may have led to a demand for harbours with deeper water. More elaborate quaysides and harbour installations would also have been needed to accommodate the deeper draught of these vessels. In the pre-conquest period a shallow shelving beach or 'hard' seems to have been the only anchorage necessary for the almost flat bottomed ships of the period (Fenwick 1978, 179-92; Hobley & Schofield 1977, 37) and this could easily have been provided by riverine ports.

The totally new foundation of many coastal ports from the eleventh century onwards (Newcastle, Hull, Boston, King's Lynn and Great Yarmouth for example) lends support to the theory of increasing trade and the enlargement of cargo vessels, but other factors may have played a part, notably the availability of land and the deliberate action taken by the lord of the land to found a market there. On the other hand, timber or stone waterfronts have been discovered by excavation at a number of later medieval ports including King's Lynn (Clarke & Carter 1977, 100-11, 413-23), London (Hobley & Schofield 1977, 37-44; Milne & Milne 1978), Hull (Whitwell, personal communication) and Harwich (Bassett, personal communication), and the remains of ships have also been discovered in London (p. 83).

Historical evidence for international trade has been briefly discussed above. Comparable archaeological material will be presented below, using the excavations in King's Lynn as the source of evidence. King's Lynn will also be taken as the example for a brief discussion of port development in the middle ages.

The foregoing pages have mentioned King's Lynn as a post-conquest east coast port, situated on the east bank of the River Great Ouse only 6 km from the Wash. With its geographical position and harbour installations it conforms to the normal post-conquest ports mentioned above and from the twelfth century onwards it participated in the medieval trade of the east coast ports. Throughout the middle ages it maintained close contacts with Norway, and of all English ports it was probably the one with the closest and most long-lasting links with the Baltic and the Hanse traders (Carus-Wilson 1973; Purcell 1973).

Most of the objects traded were of a perishable nature such as those mentioned above and are, therefore, unlikely to have been preserved except under very favourable conditions. Our main archaeological source of information comes from the pottery which has been discovered from excavations in the town. This consists both of locally made wares and wares acquired from elsewhere in England or abroad.

The character of local pottery is known as Grimston ware, made in kilns at Grimston about 8 km east of King's Lynn. A grey, sandy, reduced ware, it occurs in the form of mainly unglazed cooking pots and bowls, and olive-green glazed jugs, some highly decorated (Clarke & Carter 1977a, 191-210). The jugs in particular are sufficiently distinctive for them to have been identified in archaeological contexts abroad. Imported wares are readily distinguishable from native Grimston ware and Fig. 9.2 shows the distribution of the sources of foreign pottery found in King's Lynn as well as Grimston ware found on the Continent.

The earliest imported wares found in King's Lynn consist of a single sherd of a Badorf-type Reliefband amphora (which could date from the tenth century but which is probably twelfth or early thirteenth century in date), late twelfth or early thirteenth-century Rhenish Pingsdorf-type jars and blue-grey handled ladles, and north French glazed jugs.

The century from c. 1250 saw the stonewares of Seigburg and Brunssum replacing the earlier Rhenish material and dominating the Low Countries pottery which is otherwise represented by a few sherds of Aardenburg ware. North French wares continued to be imported, but were supplemented by pottery from the Bordeaux region. From 1350-1500 the imported wares were predominantly Rhenish stonewares—Siegburg and Langerwehe— with only a small quantity of French material.

The fact that a distinctive type of pottery was made at Grimston and distributed through King's Lynn adds another dimension to the archaeological material. Grimston ware has been discovered on a number of thirteenth and fourteenth-century sites on the Continent, the material mainly being found during excavations of medieval ports.

A complete Grimston-ware jug has been discovered at Aardenburg, Netherlands, sherds found at Ribe in Denmark, at Gamla Lödöse near Gothenburg in Sweden and at Oslo, Trondheim and Bergen in Norway. The predominantly coastal distribution of Grimston ware on the Continent is not surprising, all the sites where it has been found were ports of considerable importance in the medieval period (Clarke & Carter 1977, 447).

The imports and exports of pottery from King's Lynn reflect some aspects of Lynn's overseas trade in the middle ages. Contacts with the Low Countries, for instance, continued throughout the middle ages; north France was superseded by Bordeaux as the main source for French pottery in the

Fig. 9.2 The sources of goods imported into King's Lynn and Continental sites where Grimston ware has been found

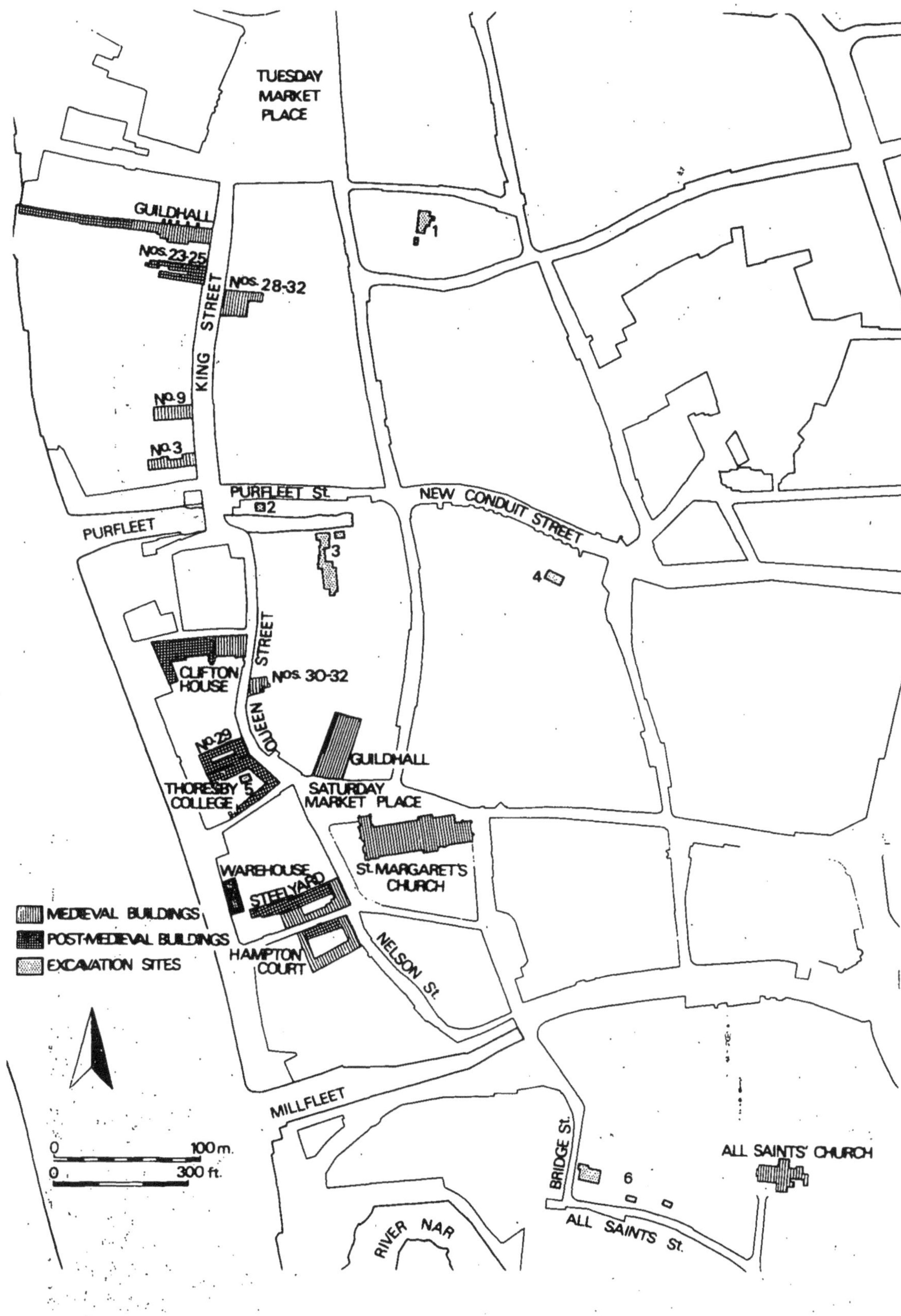

Fig. 9.3 The waterfront at King's Lynn showing present medieval and post-medieval buildings

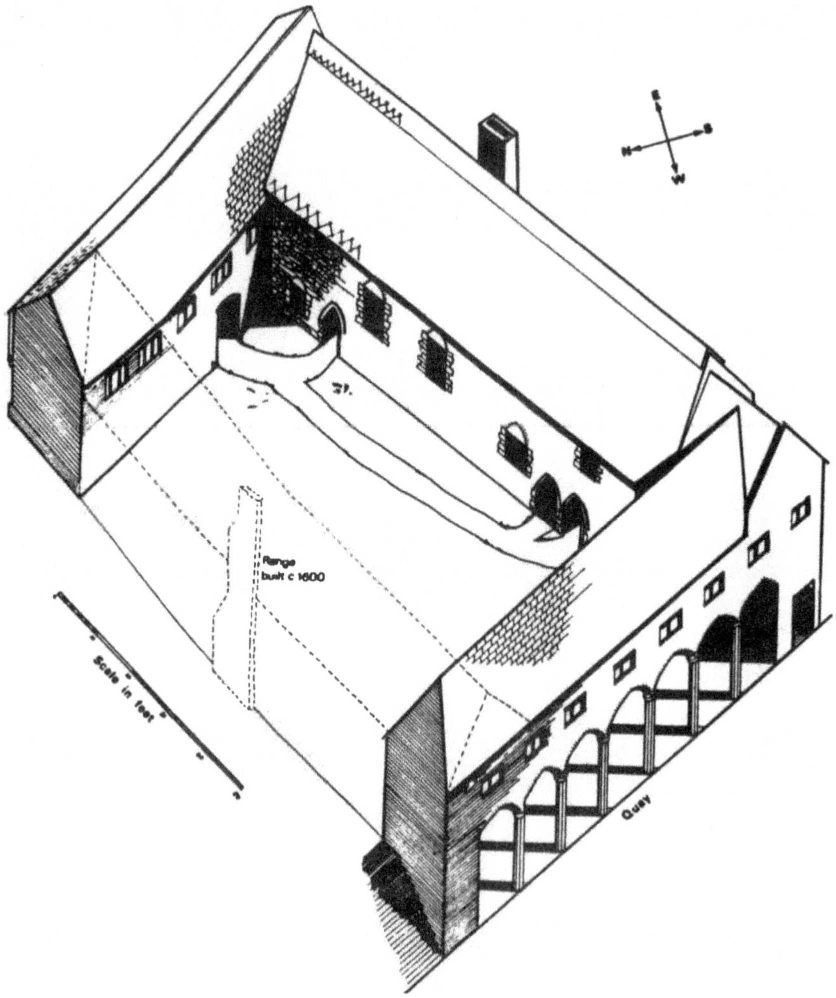

Fig. 9.4 Hampton Court, King's Lynn. Reconstruction. After Parker (n.d.)

Fig. 9.5 Hampton Court, King's Lynn. The warehouse-wing modified for modern use. Photograph: Helen Clarke

middle of the thirteenth century; Scandinavia received Grimston ware in the late thirteenth and fourteenth centuries. There is, however, a very considerable gap in the archaeological record. Historical documents suggest that there was trade between Lynn and the Baltic throughout the fifteenth century but we have no archaeological evidence of this. No Grimston ware has yet been discovered in the Baltic ports, and no southern Scandinavian pottery (either from the Baltic coasts or the Danish islands) has been recognised in King's Lynn. The only archaeological indication of English-Baltic trade in the middle ages comes from Lübeck where three sherds of Yorkshire pottery and a fragment of a mortar in Dorset conglomerate have recently been excavated (Hurst 1978). The discovery of Grimston ware along the Baltic coasts may only be a matter of time.

Both archaeological excavation and architectural evidence have been used to work out the development of King's Lynn in response to its trading activities. In particular, the extant houses of medieval merchants have been investigated (Pantin 1962-3; Parker 1971) and their presence, plus evidence from documentary sources and archaeological excavations, has been used to form a picture of the waterfront area of the town at different stages of its development (Clarke & Carter 1977; Clarke forthcoming). By doing this is has been possible to show that the commercial, riverbank area of the town was deliberately extended westwards during the middle ages and that the properties on the bank of the river gradually increased in size and importance (Clarke 1973). Figure 9.3 illustrates the present waterside topography of King's Lynn. It can be seen that in most instances the medieval buildings lie with their western faces further from the river than do the post-medieval (mostly sixteenth-century) buildings. The present position of the riverbank is even further west.

Several of the waterside buildings show interesting architectural features indicative of their wealth in the middle ages. Clifton House possesses a fine fourteenth-century brick undercroft (Pantin 1962-3, 175-7) and two fourteenth-century tile pavements (Eames 1955); the Steelyard, the only surviving building in England of the German Hanse, is a fifteenth-century brick and timber structure occupying a large site next to the river; and, perhaps most interesting of all, fifteenth-century Hampton Court possesses a warehouse-wing lying parallel to the river. Figure 9.4 shows the suggested reconstruction for this unusual building and Fig. 9.5 indicates the present modified state of its ground floor. This is the only arcaded medieval warehouse at present known in England, although the foundations of a very similar structure have been excavated at Baynard's Castle, London (Marsden, personal communication). John Schofield has also discerned arcaded warehouses at the Steelyard, Custom House, Queen's Hythe and Billingsgate on a sixteenth-century view of London (personal communication) and a similar building was standing until recently in Ballyglass, Northern Ireland (Delaney, personal communication). Similar buildings were known in the Rhineland and in the Hanseatic towns (Ellmers, personal communication) and it may be possible to trace links between the English and continental examples.

The extant buildings along King's Lynn's medieval waterfront give us one of the best pictures we have of what a medieval port on the east coast of England must have looked like. Archaeological and documentary study can tell us how the medieval port functioned. Further work of this type now being undertaken on other east coast ports will eventually give a similar understanding of many other medieval ports on the eastern seaboard of England.

REFERENCES

Ballard, A., 1914, The English Borough in the Twelfth Century, Cambridge

Barley, M.W., 1964, 'The medieval borough of Torksey: excavations 1960-2', Antiq.Jnl., 44:165-87

Beresford, M. and Finberg, H.P.R., 1973, English Medieval Boroughs: a Handlist, Newton Abbot

Carus-Wilson, E.M., 1962-3, 'The medieval trade of the ports of the Wash', Med. Archaeol., 6-7:182-201.

Carus-Wilson, E.M., 1973, 'Die Hanse und England', in Hanse in Europa, Cologne: 85-106

Clarke, H., 1973, 'The changing riverline of King's Lynn, Norfolk, in the middle ages', Int. Jnl. Naut. Archaeol., 2:95-106

Clarke, H., in press, 'Roman and medieval ports in north-west Europe', in V. Fenwick (ed.) Archaeology and the Sea, London

Clarke, H., forthcoming, 'The medieval waterfront of King's Lynn, Norfolk', in B. Hobley and G. Milne (eds) CBA Research Report

Clarke, H. and Carter, A., 1977, Excavations in King's Lynn 1963-1970, Society for Medieval Archaeology Monograph Series: 7

Darby, H.C., 1951, An Historical Geography of England before 1800, Cambridge

Davison, B.K., 1967, 'The late Saxon town of Thetford: an interim report on the 1964-6 excavations', Med. Archaeol., 11:189-208

Dunmore, S., et al., 1972, 'The origin and development of Ipswich: an interim report', East Anglian Archaeology, 1:57-67

Dunmore, S. and Carr, R., 1976, 'The late Saxon town of Thetford', East Anglian Archaeology, 4:1-21

Dunning, G.C., 1977, 'Mortars', in Clarke and Carter 1977

Eames, E., 1955, 'The products of a medieval tile kiln at Bawsey, King's Lynn', Antiq. Jnl., 35:162-81

Fenwick, V. (ed.), 1978, The Graveney Boat, B.A.R. 53

Hall, R.A. (ed.), 1978, Viking Age York and the North, CBA Research Report 27

Helle, K., 1967, 'Trade and shipping between Norway and England in the reign of Håkon Håkonsson (1217-63)', Sjøfartshistorisk Årbok: 7-33

Hobley, B. and Schofield, J., 1977, 'Excavations in the City of London: first interim report', Antiq. Jnl., 57:31-66

Hurst, J.G., 1978, 'Medieval and post-medieval imports of pottery at Lübeck', Lübecker Schriften zur Archäologie und Kulturgeschichte, 1:113-17

James, M.K., 1971, Studies in the Medieval Wine Trade, (ed.) E.M. Veale, Oxford

Mahany, C.M., et al., forthcoming, Excavations in Stamford, Lincs., Society for Medieval Archaeology Monograph

Milne, G. and Milne, C., 1978, 'Excavations on the Thames waterfront at Trig Lane, London, 1974-76', Med. Archaeol., 22:84-104

Nedkvitne, A., 1976, 'Handelssjøfarten mellom Norge og England i høymiddelalderen', Sjøfartshistorisk Årbok:7-252

Pantin, W.A., 1962-3, 'The merchants' houses and warehouses of King's Lynn', Med. Archaeol., 6-7:173-81

Parker, V., 1971, The Making of Kings Lynn, London and Chichester

Parker, V., n.d., Thoresby College — Hampton Court, King's Lynn

Power, E., 1941, The Wool Trade in English Medieval History, Oxford

Purcell, D., 1973, 'Der Hansische "Steelyard" in King's Lynn, Norfolk, England', in Hanse in Europa, Cologne; 109-12

Salzman, L.F., 1931, English Trade in the Middle Ages, Oxford

Sephton, J., (trans.), 1899, The Saga of King Sverri of Norway, London

van Es, W.A., 1969, 'Excavations at Dorestad; a pre-preliminary report', Berichten ROB, 19:183-207

ABBREVIATIONS

Antiq. Jnl.	The Antiquaries Journal
Berichten ROB	Berichten van de Rijksdienst voor het Oudheidkundig Bodemonderzoek
CBA	Council for British Archaeology
Int. Jnl. Naut. Archaeol.	International Journal of Nautical Archaeology and Underwater Exploration
Med. Archaeol.	Medieval Archaeology

KEYWORDS

Medieval.

Britain; east coast; King's Lynn.

Pottery; ships; waterside buildings; harbour installations.

Trade.

10. THE CAPABILITY OF THE SINGLE SQUARE SAIL RIG A TECHNICAL ASSESSMENT

Thomas Gillmer

Summary

A single square sail rig is an airfoil system, adequately elementary to be adapted to a simple theoretical analysis. There are some significant factors in such an analysis that can be isolated for special emphasis in relation to the sail's effectiveness in medieval vessels. These are discussed as:

1. Proportion and size (relative)
2. Curvature
3. Apparent wind angle
4. Wind force
5. Sail material and texture

The utilization of a dimensionless sail coefficient provides an indication of the comparative sailing potential or power factor. Such coefficient is S.A.—Displ. = $SA/\nabla^{2/3}$. Likewise, the sailing performance depends on a weight factor which can be expressed as a dimensionless number indicating the relative bulk of a vessel to be forced through the water. This is the Displacement to Length ratio $\Delta/(L \times .01)^3$. It can be concluded analytically that the single square sail rig is not in itself a limiting deterent to sailing performance, but during the middle ages, such deterent appears to have been the general state of the art. This factor, as borne out by archaeologists and historians, has evidently been a varient depending on regional culture.

To so discuss the single square sail rig it is appropriate first to circumscribe the discourse with certain limits, first perhaps in some definitive terms. It is understood that our intention is to deal with an historic, indeed, an ancient sailing rig and its potentialities. Further, it is the intention to apply some of the reasoning that has become acceptable theory in today's aerodynamic and hydrodynamic technology. It is not the intention to dwell on or even to introduce any obscure or advanced theories of dynamics or to reveal any newly discovered potential in the responses of ancient square sail driven boats.

I think that the term "square" sail is understood to be a sail of basically rectangular proportion; very simply to include all sails described as "square" it can be said that they have essentially four sides, the uppermost side being attached to some form of yard arm hung horizontally on the mast at the yard's middle. The sail is thus symmetrically distributed on both sides of the mast, more or less. In this arrangement the square sail is additionally given the means to be braced around on either side of the mast and with respect to the ship, the purpose being to facilitate a sailing direction in the sea at a variance to the direction of the wind.

The sail, in order to accomodate the basic action described, necessarily has certain rigging that is more or less universal. The supported mast has a fore and an after stay, in addition, there is generally some form of lateral staying called shrouds. This is the standard and fixed supporting structure of any single mast from antiquity until today and is mentioned here because it contributes to some of the constraints of the sails effectiveness, as will be seen.

The running rigging to control the sail generally is of two basic parts— the braces for the yard, one at each end, and the sheets from the two lower corners of the sail. Both braces and sheets run aft to some deck attachment. To this very simple arrangement has been added various other rigging of different degrees of sophistication, both for easier handling of large sail spread and for improving the sailing effectiveness of the single square sail.

With our sail now hoisted and ready for the wind let us leave it for a moment and look in a more abstract way at the theoretical action and reaction of the moving air on or across a rectangular surface of fabric. The wind striking such a surface in a perpendicular line of impact will have the simple action reaction effect of a hose stream against a wall in a much simplified analogy. The wind will deflect vertically and laterally, spilling over and around the edges. The rectangular fabric sail will tend to move away directly with the wind stream and if attached to a mast fixed to a boat in the water it will move the boat at some speed much less than the wind because of various resistances and losses of energy. This is the most simple and elementary form of sailing. We could say perhaps, too, the most primitive and unimaginative. It is not too much to speculate, because there is quite adequate evidence, that ancient sailors as early as the second millennium before Christ and likely earlier had discovered that by bracing their sail around as far as they could on one side or another, and by using a steering oar to control their course, they could sail across the wind's track with considerable advantage.

The above description of the wind striking the sail directly from aft when the sail is set across or athwartship the boat, while elementary, is a valid and justifiable way to go—if that is the direction you wish to go. It is called sailing downwind. In the present day sailing community, there are special large, lightweight sails made primarily for this purpose and held out laterally by poles, in order that they may spread the greater part of their surface to catch the overtaking air.

It is not essentially our interest to further investigate the simple techniques of downwind sailing. The significant interest is, of course, the capabilities in sailing with a square sail toward the direction of the wind.

We can see on many old documents, in drawings of the medieval period, the connection of the square yard to the mast so fitted as to indicate rotational movement from port to starboard—this attachment was for many centuries a simple yoke of rope on which heavy, round wood beads were strung, called paralls. The yards could then be braced with some facility on either tack. (Relative direction of the wind).

Now we encounter the abstract situation of the wind striking the rectangular foil from an oblique direction. The surface of the foil or sail curves slightly with the wind pressure and becomes a shaped airfoil. In this manner, we can assume the classic resultant or total force acting on the sail in a direction normal to the forward face of the sail. By a manner of scientific logic, we can break down acting wind force into 2 components, one which is useful along the ships course, and one which is not useful but must be considered. The useful force is best called the "driving force", the other can be simply called the "heeling force".

Keeping the two components now before us, it is most obvious that the concentrated attention must be given to the former without neglecting the latter. The useful driving force is dependent upon many factors. We can list some of the more significant of these and perhaps examine them more closely.

1. Sail Proportion and Relative Size
2. Sail Curvature
3. Angle to apparent wind
4. Wind Force
5. Sail material—(flax, cotton, wool, leather etc).

1. <u>Sail Proportion and Relative Size</u>

The first of the above is most elemental of course—the larger the sail the greater surface it presents and the greater the total force. This means little to us in terms of performance capabilities. We must analyze this factor more realistically in a dimensionless number that allows comparison of one ship to another regardless of size. The most acceptable such number today is the so-called sail area-displacement ratio. This is more precisely expressed as:

$$SA/\nabla^{2/3}$$

where SA = Sail Area in square feet
∇ = displacement volume in cubic feet of water of the ship

In addition to being a dimensionless parameter, the above relationship tells us abstractly a boat's ability or, more specifically, its capability under sail. In short, it is a useful parameter in today's world to measure a boat's sailing potential. In the context here it would be interesting to make some such estimates of boats of the Middle Ages and, as well as we can, evaluate them.

For example, first, let us take a well known classic of the Scandinavian region 1000 years ago. The Gokstad ship, being nearly intact as found, her dimensions are most reliable. She is close to 65 feet on the water, 17 ft. beam (15 ft. at the waterline). She can be rather well measured and said to have a working load displacement of close to 21.3 tons. Her main yard measures 41 feet[1] and consequently her single sail could have been nearly 40 ft. width. The sails height must be a slight speculative, but not much. The mast of this ship was not intact, but judging from its diameter, it must have been at least 35 feet from deck to halyard sheeve. This is, to a large extent, substantiated by comparison with other boats of the same type and period. A sail on such a mast and with a 41 foot yard, would have been approximately 25 feet head to foot.

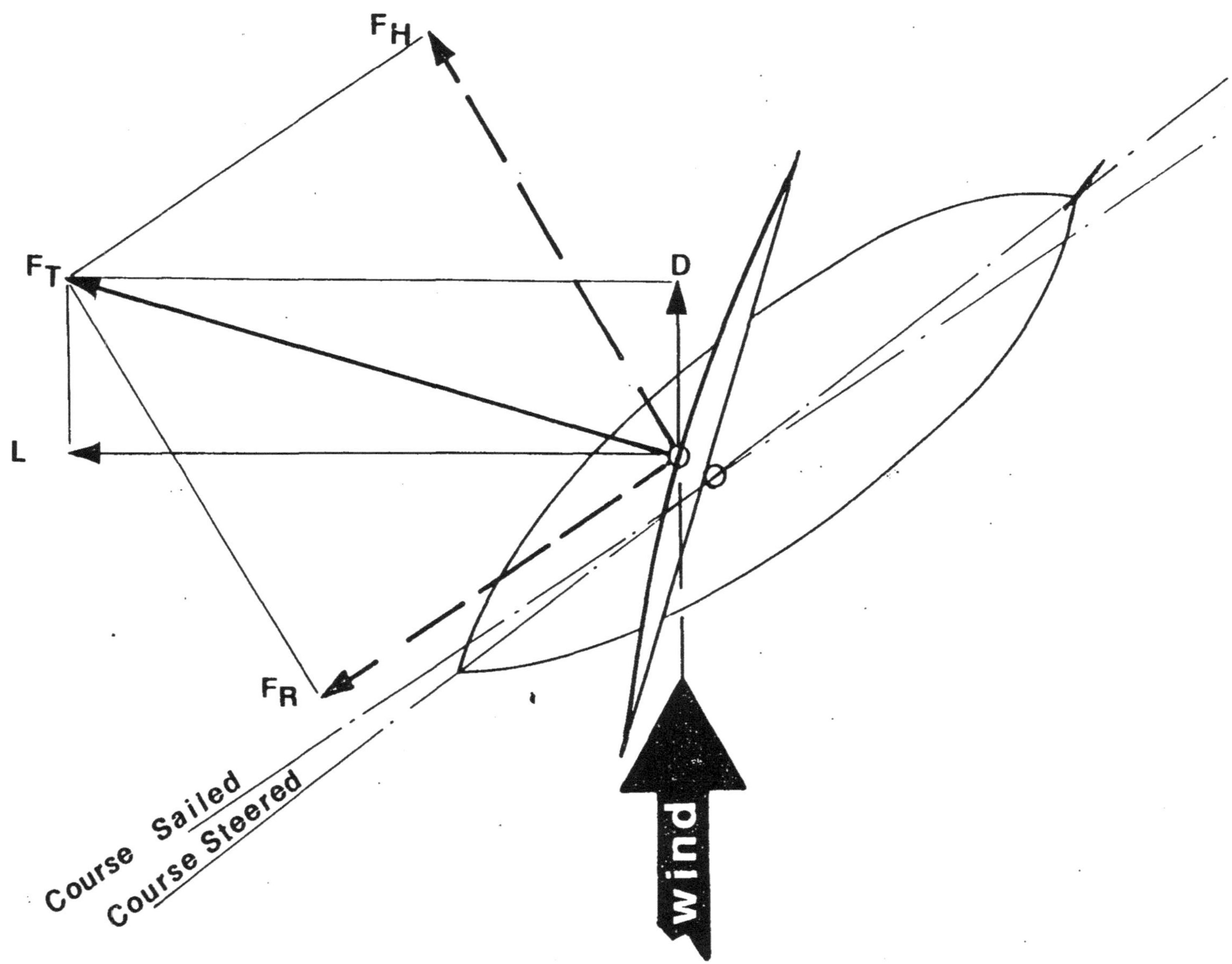

F_H
F_T
D
L
F_R
Course Sailed
Course Steered
wind

Fig. 10.1 This diagram shows the essential forces and their components which produce the mobility of a ship under a square sail.

The apparent wind striking the sail from a direction on the port bow creates a resultant total force (F_T), This force is made up of the two components related to the ship, the useful driving force (F_R) and the heeling force (F_H).

With relation to the sail's effectiveness only the total force, F_T, is usually separated for analysis into the lift component L which is perpendicular to the wind's direction and the drag component D which is the wind's direction.

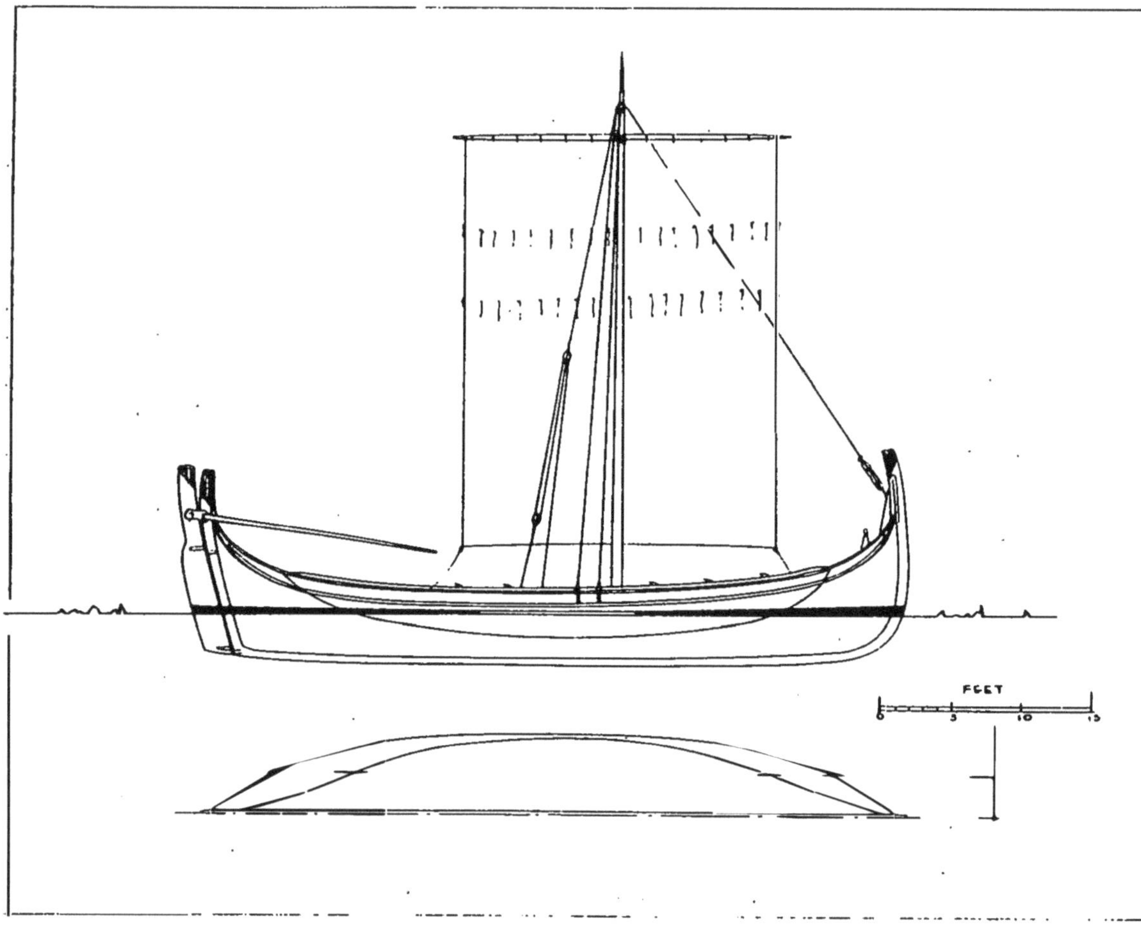

Fig. 10. 2 This Nordland's boat of the North west coast of Norway until recent times, is an excellent example of a surviving single square rigged boat—a direct heritage from the Middle Ages.

Trials on one of these boats in 1974 by the Danish National Museum showed her to be extremely fast and weatherly. This is essentially the result of her efficient sail form of good aspect ratio, and the ability to brace the yard up to a sharp angle as well as a fine, light sailing hull.

Using these dimensions, her sail area-displacement ratio is thus 13.2. The figure itself is of value only in relation to that of other sailing vessels. For example, this figure, in comparison to modern sailing yachts, is interesting. A cross section taken recently and published by a nautical journal in the U. S.[6] and listing key parameters of current popular sailing yachts from 30 to 60 feet showed the Sail Area-Displacement ratios varying from 12.4 to 22.8. Thus, the Viking ship from Gokstad being 13.2, appears to be within the same range of potential sailing performance with relation to her hull volume and sail area as the modern sailing yacht. We can say here, also, that this same statement cannot be made about many slow heavy cargo ships and ships of war that existed in the centuries between; or the best of the Renaissance and the ages of exploration into the early 19th century. These great ships did not reflect good sailing potential in their dimensions.

Returning to our examination of the capability of the single square sail toward the wind our second item is:

2. Sail Curvature

The sail, when it is not being a simple projected area and an obstruction to the wind's progress but rather in its true purpose, must curve to the wind's pressure into a three dimensional shape that creates an airfoil of some effect. To describe an effective airfoil shape is difficult but, for a sail, it must to a large degree be controlled, first by the sail maker in his skills to give it draft and smoothly roped edges, flat seams, rat tailing and the rest. Next, because of its variation in changing weather and modes of sailing, reaching and hard to the wind, the sailor must control the final optimum effective curve.

The harder the wind's force (higher velocity) then the flatter the curve should be, and contrarily the lighter the wind the fuller set curvature becomes most effective. The ancient sailors most certainly learned these things by sailing and not from books—there was no need to know why.

In any case, the flow of air across the sail's surface creates a phenomenon, setting up differential pressures on the two surfaces, front and back. These differential pressures, positive on the face which the wind strikes and negative on the opposite face, in effect combine in a total pressure which is resolved, as indicated previously, to be utilized in the direction of sailing.

Referring only to the sail's forces, we can speak only of the useful resultant normally called "lift" on the forward face and the non-useful resultant simply called the "drag".

The problem is to minimize the drag and at the same time maximize the lift.

The lift in this situation, when sailing with the wind abeam or as much forward of it as the boat will allow, is the most important of all of the components of the total force on the sail. It is the pull and tug on the upwind side of the sail that literally draws the boat along and allows, if other conditions agree, the boat to go to windward.

3. The Angle of the Apparent Wind

We must understand that the wind we are speaking of on the boat's sail is most truly called the apparent wind or relative wind. It is thus understood to be measured with relation to the boat so that its angle to the sail and the sail's angle to the boat all are real measurable angles. Likewise, its velocity and the direction and magnitude of the forces on the sail are all capable of precise evaluation.

The apparent wind, then is the wind that one experiences on board the sailing vessel moving through the water (and air). It is the combination of the boat's motion and the wind's motion. So, then the direction of the apparent wind as it moves across the sail makes an angle with the sail or the line of the sail between leading and trailing edge. This angle is relatively small, but not too small when approaching maximum lift. The greater the wind's force, the smaller this angle can be to produce an effective pull, and the flatter the sail must be. It is evident from the old records and reliable few illustrations that in the middle ages, there were introduced tack lines and bunt line to help flatten and control the leading edge of the sail as well as its foot. Also, there is evidence that on the earlier craft, before the masts grew taller and heavier, the shrouds or lateral stays were fewer and also more mobile with running blocks, instead of later deadeyes. This, of course, made possible the sharper set of the sail by bracing up the yard at a greater angle. These things together with a good sharp and not too heavy hull leave little question of the boat's ability to make some progress to windward.

The apparent wind angle for optimum lift will occur in theory at something between 12° and 20°. At angles smaller, the resultant and useful forces rapidly fall off to nothing. There are several reasons for the variant angles for optimal propulsive components. First, as mentioned, wind velocity: normally, the higher the velocity the greater sustained useful force component at a smaller flatter angle. Secondly, external interference, such as hull oscillation, surrounding rigging causing turbulance. Third, sail surface roughness. There are others, but perhaps the above are most meaningful.

4. The Wind Force

There are many factors to effective sailing beyond the shape of a sail, rectangular (square) or triangular or quadrilateral. The very first confrontation facing any sailor about to push off is force of the wind. Aside from the sea it may be kicking up and the possible adverse direction it may be blowing, the simple strength of its force is a formidable consideration. As a simple matter of geophysical limitations, the winter winds of the North Sea are assumed to have been too strong for normal sailing commerce during the Middle Ages. This is a most reasonable assumption with much substantiation.

We have noted that the angle to the wind that a sail can be brought to is a function of wind force among other things. Putting aside the propulsive and drag components of the wind vectors let us return to the heeling force mentioned earlier. This force component on the sail is obviously tending to overturn or at least push the vessel down on its side using the mast as a lever. This tendency is undeniable and warrants some understanding. The heeling force is the lateral (transverse) wind component acting at the sail's centre of effort (the geometrical centre of a rectangular sail). This centre is at some distance above the water, obviously and is resisted in its lateral movement by the lateral water resistance on the hull. Now these parallel forces acting in opposition and at a distance form a couple or a tendency toward rotation. Every ship that floats upright with normal tendency to remain so has a potential righting moment that immediately develops upon any rotation about its axis. In other words, this wind and water couple causing an inclination of the ship is resisted by an opposing righting couple. This is largely a factor of the ship's beam and hull form, in addition to its weight and favorable position of centre of gravity.

The existing evidence of systemetized ballasting in vessels of the Middle Ages is inconsistant and it must be assumed that the practice was not very well understood nor was it done in an orderly way. This factor alone, probably was responsible for the low sail and mast plan of the vessels throughout the period and before.

This lack of understanding does nothing to the high probability that some vessels were adequately stable and could consequently under either temporary or fairly consistant circumstances carry more sail and for a longer duration in increasing winds. Better hull configuration and/or stowage of cargo together with high density type of cargo, accounted for this.

Low centre of gravity produced by low weights is the result of ballasting either intentionally or unintentially. This increases the righting arm and righting moment in as much as the centre of gravity is the terminal point of the righting arm. A vessel's greater beam increases the righting arm also, but often at the expense of poor sailing form in a hull too fat and full. This is a balance to be achieved by experienced shipwrights.

The matter of stability of ships, whether ancient or modern is a classic topic and cannot be more than introduced in its significant bearing on the sailing capability. It should and must be said that where the ship's stability is doubtful and minimal and is not controlled by an informed programme for ballasting, as was the case during this period of the middle ages, it is quite logical to conclude that a single square sail was the better of the several types or shapes of sails then or even now in use. It produces maximum aerodynamic response at the lowest centre of sail area.

The exception could be cited that in climates where light winds and calm seas prevailed it would have been in the interest of better performance to use loftier, narrower sails such as the Mediterranean lateen.

5. Sail Material

The substance of a sail is as important nearly as its shape when set and drawing. It is best to illustrate this matter in the present day context. Sails today for sailing yachts are made from a plastic, woven cloth called "dacron" in America "terelene" in England and so on. It is a fine, smooth plastic surface when, without wrinkles it produces a near perfect air flow about it. Compared to the previously used long staple Egyptian cotton sail cloth, "dacron" produces up to 45% higher lift component on test airfoils in wind tunnel tests. This difference is caused simply from the weave of the natural cotton fibers producing a soft but perceptable fuzz on the surface. This surface "roughness" as it is called technically causes the airstream flowing over it to break its smooth flat laminer flow and dissolve into a system of turbulence that grows, deflecting the airstream. The airflow thus disturbed is to a considerable extent less effective. (The introduction of "dacron" sails to sailing in the past two or three decades has been one of the greatest factors improving the technology of sailing).

Consider then the differences between the sail cloth fabrics of the Middle Ages and those of our best cotton sail cloth of the century past. Actually, there was probably more difference between individual vessel's sails of the Middle Ages than between their better fabrics and the 19th century's cotton sails. For an example, flax sail cloth was not unlikely in the Middle Ages and the weave of flax sail cloth produces less surface difficulty than many heavy cottons. The greater difficulty is its flexibility and tendency to stretch. This fault is not generally as great a disadvantage in a square sail as in a fore and aft sail that must snuggly fit on two or three spars at once instead of one. Probably more significant material quality factors in ancient sails that would mar their aerodynamic performance would be the hand worked seams and rough or frayed edging. The reef points, reef bands and bonnet connections, all of these have affected the optimum performance of sails in more ages of history than the middle one.

I believe that the fine weave and construction of a sail is far more important when the sail cloth is of woven natural fibre. A sail should be made so that it will assume a nicely curved foil surface that by adjustment of tack and clew lines, luff and leech lines or other jack lines can be flattened or loosened to full curve. This surface should not allow for stretch wrinkles, sag wrinkles, bulges or hard spots. A well made sail is the work of a skilled craftsman and the craft is an ancient one.

Putting the several factors discussed separately above into one ideal or optimum performance situation let us see what we may expect.

The proportion of a square or rectangular sail, first of all, should be such that it accounts for the inevitable loss or leakage of wind over the top and under foot, still retaining adequate area between for efficient cross flow. This means a sail that is substantially taller than it is wide. This relationship of height to width is referred to as the aspect ratio and is expressed for rectangular sails simply as the height divided by the sail's mean width. For a sail which is exactly square, the aspect ratio is exactly 1.0. For a low wide sail the aspect ratio might be 0.6 or 0.5 or even less. But, for a really efficient "square" sail where we have some expectations of windward capability the aspect ratio should be at least 1.5 or better.

The loftier the sail, of course, means there must be some provision for stability to counter the heeling moment. This is assured in two ways, namely adequate ballast or adequate beam (breadth of hull). In a commercial vessel the ballast is often in the form of relatively high density cargo. The fullness of hull in the middle ages, as well as before and after, was well understood in gaining cargo capacity. It, unfortunately, inhibits good hydrodynamics generally. There seems to be no good reason that sails on the medieval boats might not have taken more advantage of better aspect ratio.

The sail curvature can be controlled and evidently was in many boats of the Middle Ages when they were straining for that last degree or so of upwind slant. It is a natural and instinctive thing for a sailor to do. He does not have to be told to harden down a sheet, he does not have to understand that the air flow across the sail more smoothly improving his upwind vector, he simply feels the boat move out, responding to the sail trimming. Additional adjusting lines and ropes have been worked into sails as evidenced by early illustrations which makes the efforts of medieval sailors to go to weather quite evident.

The angle to the apparent wind was limited in a square rig by the angle to which the yard arm could be rotated on the mast. This, of course, is the obvious difference between and the advantage of the fore and aft rig compared to the square rig. This difference was minimized by boats that retained few shrouds and particularly those with running shrouds. Until the ships became larger and the masts heavier with greater sail the shrouds were comparitively simple and gave the optimum yard freedom. Such simple rigs generally prevailed during all the middle ages. Hence, their capability in this respect was as good or better than that of vessels of much later periods.

We see a vessel's capability in heavy weather as a function of its stability both axialy and directionally. The first of these reflects its beam, depth and ballasting. The low sail area of the single square sail rig seems simply to say "we cannot rely on our vessel to carry more or loftier sail". This is quite probably the determining factor of the seasonal ebb and flow of commerce and economy of the North Sea countries during the medieval period when in the heavy force of winter winds the sails required reefing to the point of nearly total loss of directional control.

We see the probable use of flax sail cloth as an effective square sail fabric. A possible cloth with a combination or mixture of flax with wool for strength and durability should not altogether be dismissed. There is the greatest of reasons as still exists among wise and conscientous sailors to use the sort of sail which sails the boat the best and not the sails which can be most cheaply made. A ship's captain knows that the better sail is the more profitable one in the long run—it will get his ship better to windward (if it will go at all) and land his cargo in his destined port at a better price. It would have been a dull seaman who did not relate some of this ability to a finely made sail.

We see thus that a "square" sail which was the mode of the medieval sailor's ship was not entirely to be faulted as a sail. It had, after all, been the only genre of sail since boats first sailed—one which has endured longer than any other rig. The mediocrity of medieval sailing was not particularly the fault of the sail, its single mast nor its four cornered shape. We see that it is very much more effective when it is 2 times taller that it is wide but the lack of consistent or controlled stability of the medieval hulls did not generally permit this. It is more effective with a smooth plastic surface but the technology of the woven fabric in the middle ages could not provide it.

And finally, a sail with its mast can never be separated from the hull that it drives. The sailing rig and the hull must be matched, it is acknowledged. This statement is even more applicable today in our renovation of sailing technology with all of its components. In the Middle Ages, the awareness of it probably would have meant little to the shipwright. His sailing rig was simple and much the same, on every ship. However, the greater variance in sailing was in the ship—the weight and configuration of its hull. A single square sail rig can be most capable on one hull and not another. As can be seen in returning to the ship cited earlier—the Gokstad Viking ship. Such a ship is capable of good sailing because of its excellent sailing hull shape and its light weight.

It is probable that ships grew in capacity, responding to the expansion of commerce. more rapidly than the technology of construction of good sailing hulls. The ship builders of the late medieval period were producing deeper, wider and consequently heavier ships without increasing by the proper dimension the length on the water nor the sail area. It is easy to see that the increase in beam and depth by themselves produces a wider and a deeper hold and that logically an increase in length would increase the amount of it longitudinally. So why were they not so extended? The reply to that must be that the technology to increase the necessary longitudinal strength was not yet sufficient. But this two dimensional increase, breadth and depth, simply compounded the difficulty of providing an acceptable hull form for sailing.

There is a very simple coefficient in the shipbuilder's art which he uses to measure the basic capability of a ship against its employment, cost, performance etc. This is called its Displacement—Length Ratio. It is obtained by dividing the Displacement (tons) by the waterline length x .01 cubed (The length is expressed in feet in the English dimensional system).[2] The Gokstad ship for example would, with its displacement of 21.3 tons and waterline length of 65 feet, measure its displacement-length ratio thus:

$$\Delta/(L \times .01)^3 = {}^{21.3}/.65^3 = 77.6$$

The result, just less than 78, is purely relative but it is a measure capable of being compared to that of any other ship in terms that are meaningful. It is, first of all, revealing here in describing a boat (in contemporary 20th century technology) as being in the ultra light displacement category. This categorical assignment with such a specific designator indicates a hull that is very easily and rapidly moved under sail (other factors of its shape being reasonable and conventional, as here). This value shows a lighter hull than a modern ocean racing sailboat. The Viking ship's hull is not ballasted of course but it is calculated with a normal sailing burden and the displacement is based upon the immersed volume of the hull.

Using another ship of the middle ages, not yet published in detail, let us take the Bremen Cog.[3] The Cog's dimensions are taken to be approximately $62\frac{1}{2}$ feet long on the waterline, 77 feet in length, and 20 feet breadth on the water at a $6\frac{1}{2}$ feet loaded draft. These immersed dimensions would provide on the basis of reasonable hull form coefficients ($C_m = .80$ and $C_p = .68$) a displacement of 126 long tons. Using this weight we find the Cog with a displacement to length ratio of 516. This compared to the Viking Ship's 78 does not say much in an encouraging way about her capability for sailing well. We must, however, quickly say after such a comparison that her stability would be much greater and thus would be her ability to carry a larger sail area. Without any evidence of the Cog's sail dimensions we are forced to abandon any further numerical analysis to speculation.

Deliberately, the example taken above shows two vessels of extremely different characteristics. At the same time their purposes are quite different as well. The purpose of the simple contrast, of course, is to illuminate the vast difference of the hull's hydrodynamic ability between two types with essentially the same square sail rig.

In summary, it might be excusable on the basis of the foregoing to postulate upon the windward ability on a medieval boat of optimum hull and rig.

The heritage of the light and strong Scandinavian hulls that perhaps reached their plateau of excellence sometime before the end of the first millennium A.D. prevailed in the construction of other Northern European craft. Such were perhaps most often found in fishing and transportation employment or other industry where growth of cargo was by its nature minimized. At any rate with hulls whose depths were of these smaller proportions, the length to beam ratios were near 4.0 and the displacement to length ratios were no more than 150. It is quite probable that such vessels could work reasonably well upwind under their single square sails. Further, if their sails were of reasonably flat weave of flax or similar sail cloth, it is possible to estimate their directions to within 55° of the true wind under good sailing conditions. This I think was their optimum and the estimate is perhaps conservative.

Other vessels of fuller hulls, greater depths, of increased draft and with high free-board windage, their bottoms with little keel showing and flat floors—such vessels could be expected to perform poorly in any direction approaching the wind. Such standard carriers of heavy trade could feel satisfaction when making any progress with the wind abeam.

The next milestone in improved sailing ship capability was the recognition of ballasting, with the accompanying increase in number of masts and sails as the middle ages were ended.

NOTES

1. The estimate of 41 ft. for the yard of the Gokstad ship is based on the associated finds; see Reference 7. The values used for computing displacement were taken from Reference 4. The conclusions drawn from the numerical examples would only be affected by substantial variations from these values.

2. The Displacement/Length ratio is similar to the dimentionless Volumetric Coefficient:

$$\frac{\nabla}{(L_{WL})^3} = \frac{\text{displacement volume (m}^3\text{)}}{(\text{length of waterline (m)})^3}$$

See Reference 8.

3. The values for computing the displacement of the Bremen Cog were taken from Reference 4.

REFERENCES

1. Bass, George F., 1972, A History of Seafaring Thames & Hudson Ltd.

2. Gillmer, Thomas C., 1975. Modern Ship Design U.S. Naval Institute Press.

3. Gillmer, Thomas C., 1972. Working Watercraft International Marine Publishing Co.

4. Greenhill, Basil, 1976. Archaeology of the Boat Wesleyan University Press.

5. Marchaj, C. A., 1964. Sailing Theory and Practice Dodd, Mead & Co.

6. Perry, Robt. H., 1977. Sailing Designs Port Publications, Inc.

7. Almgren, B., 1966. The Viking Cagner & Co. Gothenburg

8. McGrail, S. and Corlett, C., 1977. 'High speed capabilities of ancient boats'. IJNA, 6:352-3.

APPENDIX

Braces

In the discussion at the Symposium it was suggested that a sail of aspect ratio 1.5 to 2.0 would require good braces. The author agrees and ventures to add that in his experience with universally undependable sailing conditions this observation applies to running rigging in all sailing vessels. However, the observation of better sailing performance with higher aspect sails, i.e. 1.5-2.0 was not intended to substantiate any historical consideration. It is simply affirmation of a physical truth and believed to be well within the realm of practicability.

Windward Ability

It was further suggested at the Symposium that windward ability in medieval vessels may not have been top priority and that the medieval sailors were willing to wait for a fair wind. Might it not be reasoned that many sailors were strongly motivated toward shores that lay most frequently upwind. It would be unusual indeed if the ambitions of the merchants of the Hanseatic League did not suggest the advantages of sailing in the direction where their fortunes were, regardless of wind direction. There is also the point that the simple avoidance of the lee shore would be motivation enough to lend impetus to construction of medieval vessels of better weatherly performance.

The discussion further raised the point that windward ability is a function of hull design as well as sail design. This point is emphasised in the paper in comparing the displacement-length ratio values of the Gokstad ship and the Bremen Cog as an indication of sailing capability with the same sail form. In response to the comment of wetness in windward sailing,it is pointed out that this characteristic is a secondary factor in the boat's and sail's capability. The windward sailing boat that is excessively wet will perhaps ship water. If it is decked it should drain off; if open, the crew will resort to bailing. The endurance of the boat and crew on a beat to windward is both a variable and relative matter to be dealt with on another level.

KEYWORDS

Medieval.
Northern Europe.
Boats and ships; square sail.
Sail construction; sail shape; wind force; vector diagram; hull effects; parametric comparisons.

11. VIKING AGE RIGGING, A SURVEY OF SOURCES AND THEORIES

Arne Emil Christensen

Summary

A brief survey of the source material for Viking Age rigging is presented, comprising archaeological finds, iconographic details, medieval written sources and some recent comparative material. An assessment of several authors' original contributions to the question is offered; along with this is the suggestion, in conclusion, that two types of sail may be represented in the various sources.

A considerable amount of research has gone into the problem of how the Vikings rigged their ships and handled them under sail. Opinions differ widely, from claims that the Viking vessels sailed better than practically anything else afloat before or after their times, to a wholesale downgrading of the sailing, and especially tacking, abilities of their ships.

However, when one looks closer at the literature on the subject, it becomes clear that a number of the papers in the discussion, and most of the "ship chapters" found in books on the Viking age, are based on secondary sources. Accordingly, I have found it fruitful to present a brief general survey of the source material. Even in the short form necessary for a meeting paper, it will be seen that sources are in reality meagre. We have little exact knowledge, apart from broad generalisations, the rest is guesswork, more or less based on the source material. I will end the paper with a brief survey of some of the main theories, most of which have been published in the Scandinavian languages. I must stress that this is a survey, where no new material will be presented.

Geographically, the survey is restricted to what we might call "the Viking world"; Denmark, Sweden, Norway, and those parts of the Baltic and Western Europe which felt strong influence from Viking Age Scandinavia.

Chronologically, I will restrict myself to the Viking age proper, with occasional glances at comparative material from the Middle Ages and later.

The sources for knowledge of Viking age rigging fall into four groups.

1. Archaeological finds of rigging details that can be identified as such, and traces of rigging on preserved ships, including the mast support. Such finds are few from the Viking age, and some of them are difficult to interpret.

2. Iconographic material from the Viking age and the Early Middle Ages. This is numerous, but the representations are often very sketchy and must be used with much caution and proper source criticism.

Fig. 11.1 The Gokstad ship, looking forward.
Courtesy Universitetets Oldsaksamling

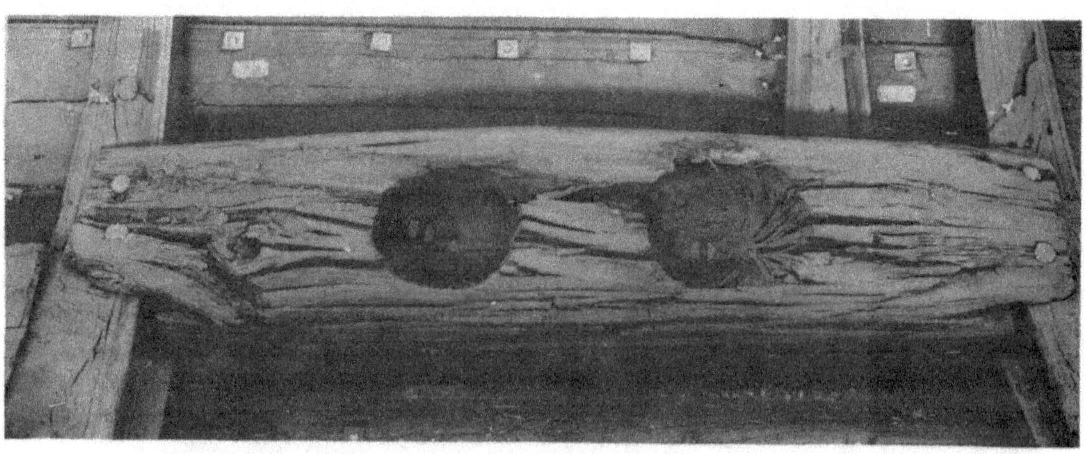

Fig. 11.2 Support for tacking boom? Gokstad ship.
Courtesy Universitetets Oldsaksamling

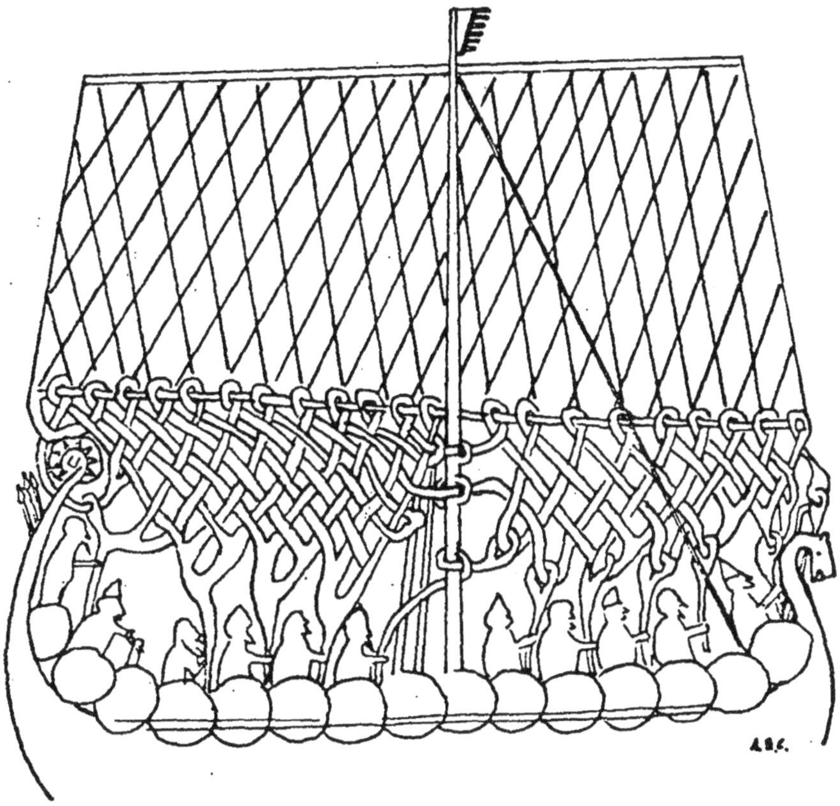

Fig. 11.3 Ship on stone Stenkyrka Smiss I. Redrawn from Lindqvist 1941

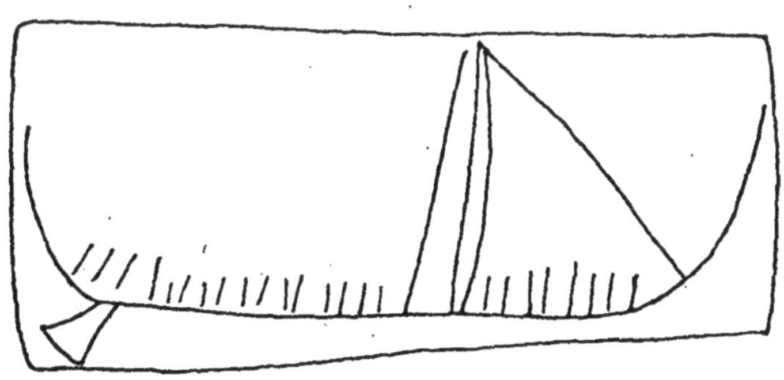

Fig. 11.4 Ship on slate from Jarlshof. Redrawn from Hamilton 1956

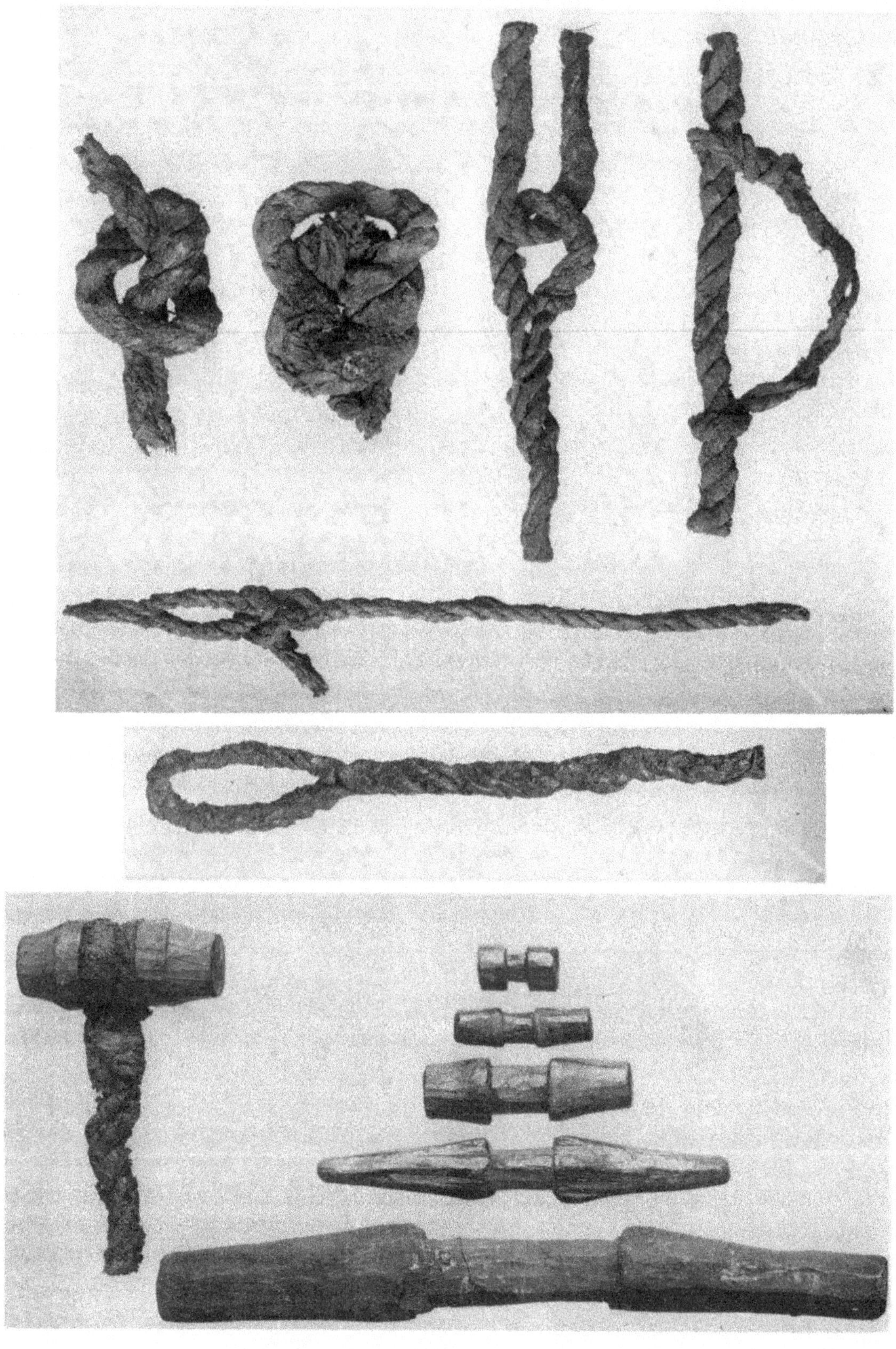

Fig. 11.5 Knotted ropes and rope-end toggles from the Oseberg ship. Not to scale. Courtesy Universitetets Oldsaksamling

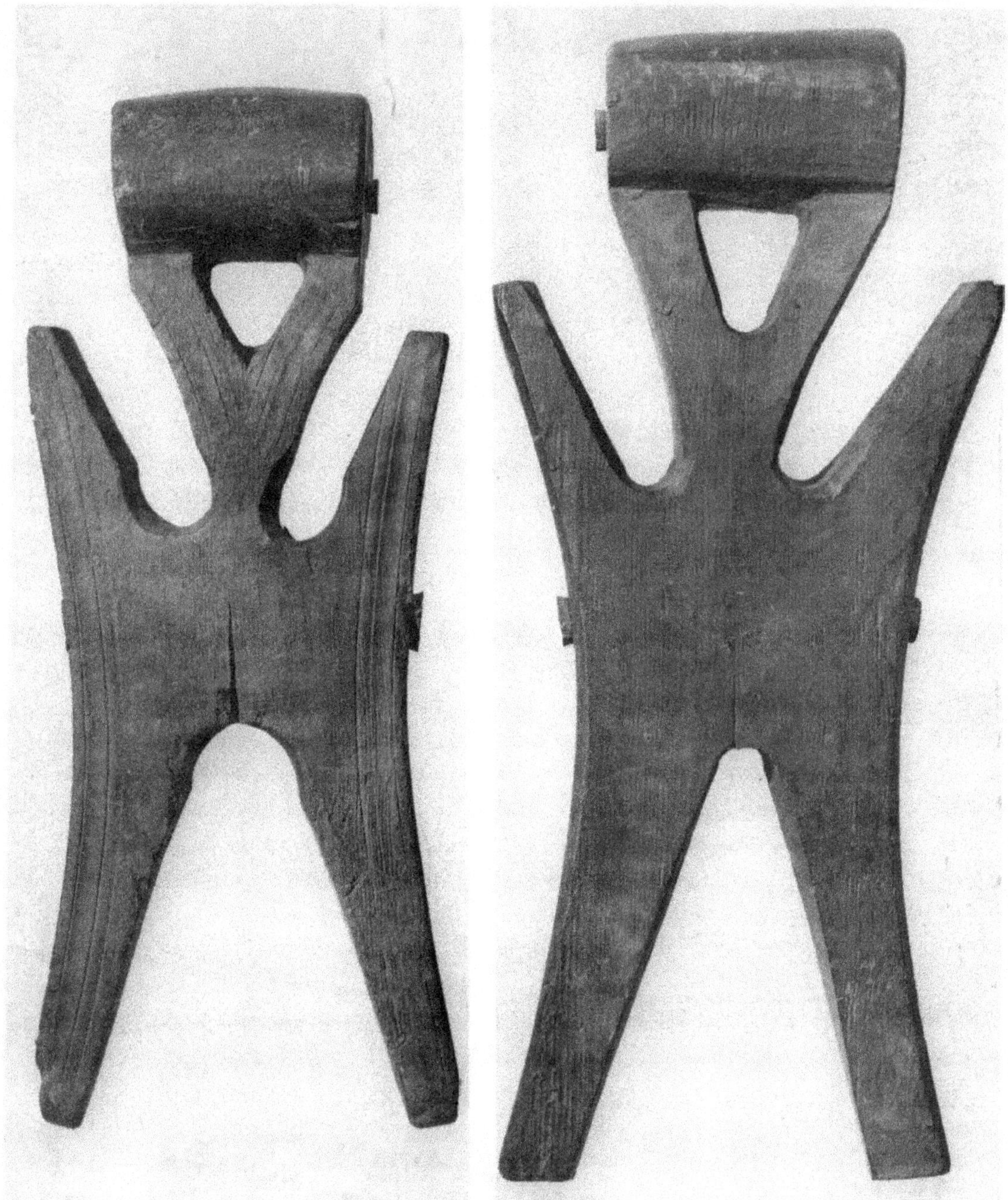

Fig. 11.6 Two of four objects from Gokstad ship, interpreted as shroud-stretchers or sheaveless blocks. Oak, the longest piece is 0.6 m long. Courtesy Universitetets Oldsaksamling

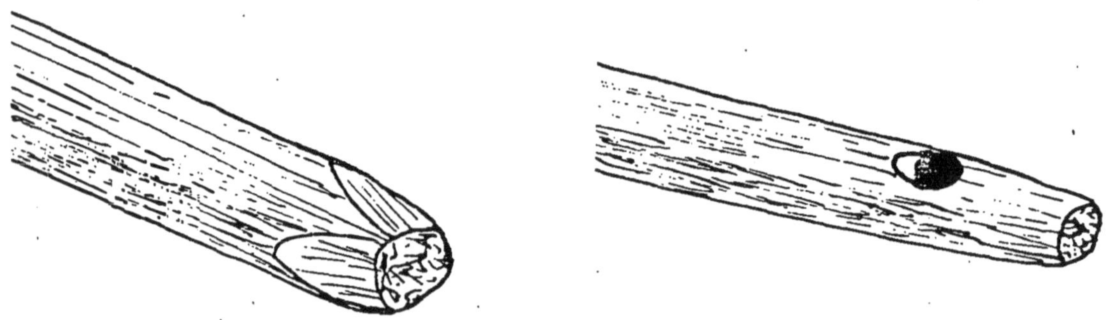

Fig. 11.7 Spar fragments from small Gokstad boats, mast heel and yard with hole for the head-earing? Sketches by author

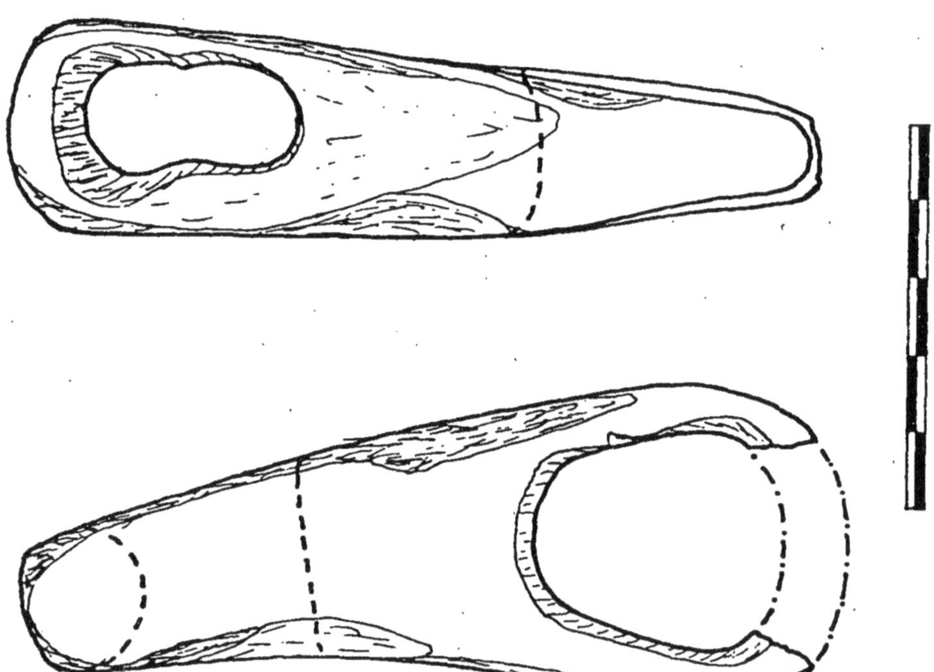

Fig. 11.8 Fairlead? of bone. Similar pieces of wood are numerous in medieval contexts, and two were found in the Gokstad ship. They were probably used in running rigging. Drawing by the author, scale 5 cm

3. Mediaeval written sources, dealing with ships and seamanship. Most of the material is in Old Norse or Icelandic: laws, poems and sagas.

4. Recent comparative material, from districts where small square-sailed boats were used until c. 1900 A.D. Most of the material comes from western and northern Norway, where extensive surveys have been carried out. Numerous boats with complete rigging exist, and it is still possible to interview fishermen who have sailed the boats in their youth. In recent years, new interest in sailing the traditional types for pleasure has given considerable experience of how they behave under sail. Replicas of Viking age vessels with reconstructed rigging have also been used in comparative experiments. They must be rated as a source equal with the recent square-rigged boats.

The sources all show that the Viking ships had one mast, placed amidships or slightly forward. Our evidence is the mast steps found in nearly all vessels. Gokstad (Nicolaysen, 1882), Oseberg (Shetelig, 1917a), Tune (Shetelig, 1917b), Äskekärr (Humbla, 1934), The five Skuldelev ships (Olsen & Crumlin-Pedersen, 1967), all have a step for the single mast. The iconographic sources also show the single mast, and the written sources support this.

The shape of the mast is well known from medieval and later evidence. Round cross-section, tapering slightly towards the top, with a thick "lump" near the top. This lump supports the standing rigging, and gives strength around the haliard hole. No mast from the Viking age exists complete, however (In the Gokstad find, the ship's mast and a mast for one of the small boats are fragmentarly preserved, both have the bottom squared for the mast step).

It has been suggested that the heavy mast partner supporting the masts of the Oseberg, Gokstad and Tune ships was so sturdy that the ships could be sailed without standing rigging (Brøgger & Shetelig, 1951). The sources speak against this. Evidence for standing rigging is found in the iconographic sources (for example: Lindqvist, 1941 vol. 1 Fig.97; Hamilton, 1956 pl. XXI Fig.1). Among the ship finds, the iron rings on the sheerstrake of the Ladby ship and the osier rings in Skuldelev 3 are the only definite evidence, both are best explained as shroud rings (Thorvildsen, 1957 Fig.22; Olsen & Crumlin-Pedersen, 1967 Fig.43).

The mast carried a single square sail spread by a yard. Yards have been found in medieval contexts in Bergen and Tønsberg, Norway (unpublished), and there were complete and fragmentary yards found with the small Gokstad boats (Nicolaysen, 1882 pl.IV nr. 12). With the yards should be mentioned the Oseberg parrel (Shetelig, 1917a Fig.121). A number of parrels have been found in the medieval excavations in Bergen, Tønsberg and Oslo, and 19th century parrels look exactly the same, so there is little doubt about the identification of the Oseberg (or the medieval) ones.

It is firmly established, then, on reliable sources, that Viking ships had one mast, a square sail hoisted on a yard, with the yard held to the mast by a parrel made from a naturally curved piece of wood.

When we go on to the sail and the running rigging, things are less certain.

The medieval written sources indicate that wool was the usual material for sails, and two 19th century woollen sails are preserved in Norwegian museums (Hordamuseet, Stend, Hordaland and De Heibergske Samlinger, Amla, Sogn). Fragments of a sail were probably found during the excavations of the Gokstad ship. In the words of the excavator: "At the fore-end of the ship was found, among other things, a bundle of woollen cloth of yellowish color, though doubtless originally white, with stripes of red cloth sewn thereon, the whole of which, presumably, had been intended for the tent, an opinion corroborated by the fact that within the bundle there were found pieces of thin hemp-rope, in all probability, the fastenings of the tent." (Nicolaysen, 1882, p. 37f.). Today, only a few small fragments of the cloth exist, so it is impossible to say anything about size or cut, but the description might as well indicate a sail, like the ones shown on the Gotland stones, as a tent. The red stripes may be indicated on the Gotland stones by the oblique lines crossing the sails, and ropes are indicated in abundance.

The evidence for running rigging is also scanty and not conclusive. Both the Oseberg and Gokstad ships have cleats aft, the Oseberg ship also forward. In the Gokstad ship the cleats are placed symmetrically, three on each side. The best explanation for the cleats is that they were for belaying running rigging, but we do not know which ropes led where. In Gokstad and Skuldelev 1, a sturdy block of wood is nailed to the ships' sides in front of the mast. In Skuldelev 1 the block has three "sockets", in Gokstad two. It has been suggested by Åkerlund (1956) that this is a support for a boom used to stretch the luff when tacking. Åkerlung combines this with the Old Norse "beitiáss", which translates literally as tacking spar.

The various written sources give some names for various ropes in the rigging, some of which are difficult to interpret, while others can be identified by their equivalents in modern Icelandic and west Norwegian dialects. The material was assembled by Falk, but he makes no distinction between early and late medieval sources, so his results should be checked against the original sources. In his paper on Viking age rigging, Åkerlund draws heavily on the work of Falk, in combination with rigging details from the Oseberg and Gokstad finds. If we turn to inconographic sources, the haliard and bowline are clearly shown on some of the Gotland stones, while some of the medieval seals and grafitti show sheets, braces and reef-points, similar to what we know from 19th century boats. The haliard hole is preserved in medieval masts (One mast and three models, Bergen, unpublished) and it is reasonable to believe that the Viking age haliard looked like what Norwegian fishermen used well into the 19th century, a single rope, well tallowed and rove through a hole at the masthead

It is a remarkable fact, that the iconographic representations from the Viking age proper do not show braces, and that the more detailed of the Gotland stones show a very complicated system of ropes under the sail instead of plain sheets.

On 19th century craft in western and northern Norway, the tack was secured through a hole in the sheerstrake forward, or made fast to a belaying pin set in the hole. Holes that may have had this use are found in Skuldelev 3, the Oseberg ship, and a boat model from Dublin. This evidence is somewhat contradictory to the Gotland stones.

Summing up, we are left with very scanty knowledge based on tangible sources. A square sail, spread by a yard on a mast amidships. The mast had shrouds, and possibly a stay; the sail was hoisted by a simple haliard and controlled by sheets (?); and the luff was stretched by a tacking boom(?) or bowline(?) or both. That is about all.

Among the authors who had discussed Viking age rigging, I will only mention a few who have made original contributions to the field. Until c. 1950, the square sail was taken for granted, but few details had been discussed. Brógger and Shetelig (1951) suggested that the Viking ships were sailed without standing rigging, but otherwise did not discuss rigging much. In 1953 Heinsius submitted a paper where he argued that the Gotland stones show square sails that were generally carried longitudinally, like a fore-and-aft sail, that the sails had a lower yard, and were controlled by a very complicated system of sheets. These theories were vividly contradicted by Kielland, (1954) who mainly used 19th century parallels from northern Norway in his argument.

In his 1956 paper, Åkerlund attempted a synthesis of iconographic, archaeological and written sources, and backed some of his theories with small-scale experiments. He suggested the use of spreader booms, both for tacking and sailing with the wind aft, and saw the system of ropes shown on the Gotland stones not as sheets, but as a system of brails for shortening the sail, like those known in the Mediterranean in classical antiquity. Experimenting in a small boat, he found that such a system of brails worked in shortening sail.

The mysterious ropes were again discussed by Almgren and Christensen (1966). It is suggested that the ropes under the sail may be a system of horizontal brails, for speedy shortening of sail in squall weather. Again, 19th century parallels from western Norway are used to explain the Viking age.

S. Haasum, in her paper on Viking age rigging and navigation (1974) compared the Viking ships to modern ocean-racing yachts, and tried to evaluate sailing capacity. The Gokstad ship, with the rigging reconstructed mainly according to Åkerlund, was taken as "type specimen". The result, apart from the obvious one, that modern yachts sail better than the Viking ships did, is a general downgrading of the sailing abilities of Viking ships. According to Haasum, tacking was next to impossible, and at any rate impractical. So far, there has been little opposition to Haasum's theories, but some of the results are based on rather uncritical use of uncertain sources.

Recently, E. Andersen has carried out systematic measurements on traditional Norwegian square-rigged boats. The regularities found will be applied to reconstructions of Viking age sails. (Andersen 1976)

The experiments carried on with the Viking ship replicas built by Danish boy scouts have not been published fully, but the ships sail well both on and by the wind. The sails used are in most cases rather small square sails, in one case a lower yard is used but no results of how this works have been published. Compared to the use of traditional square-rigged Norwegian boats, it seems definite that the Viking age hulls have good tacking potential, and it is also clear that when the right balance between hull and sail is found, the light flexisble Norse vessels show remarkable speed possibilities. So far, all experiments have been carried out with square sails of "modern" type, that is, based on 19th century models, and made from cotton or synthetic materials. It would be rewarding to continue the experiments with wool as sailcloth, and base reconstructions on the various interpretations of the Gotland stones.

The problem of Viking age rigging is far from solved, and both experiments and "writing-table" research is necessary. It may even be that we must continue work with two different concepts of Viking Age rigging. One is the square sail as we know it from the Middle Ages and until c. 1900 A.D. The other is the sail shown on the Gotland stones. It may be that the "simple" square sail is a younger form, developed during the Late Viking Age or the Early Middle Ages, and that the sail shown on the Gotland stones is an older, more complicated rig which became obsolete around 1000 A.D. If this is the case, we know even less about Viking age rigging than we thought.

BIBLIOGRAPHY

Åkerlund, H., 1956, Åss och beitiåss. Unda Maris, 1955/6. Gotenburg.

Almgren, A, (Ed.), 1966. The Viking.

Andersen, E., 1976. Hals og skaut, mast og segl. Både og rasejlsrigninger på norskekysten. Norsk Sjøfartsmuseum, Årsberetning 1975. Oslo.

Arbman, H., 1940. Der Årby-Fund. Acta Archaeologica XI Copenhagen.

Brøgger, A. W. & Shetelig, H., 1950. Vikingskipene. Oslo.

Ellmers, D., 1972. Frühmittelalterliche Handelsschiffahrt in Mittel-und Nordeuropa. Neumünster.

Falk, H., 1912. Altnordische Seewesen. Wörter und Sachen IV Heidelberg 1912.

Hamilton, J. R. C., 1956. Excavations at Jarlshof, Shetland. Edinburgh.

Heinsius, E. & P., 1953. Hvordan seilte vikingene med sine båter? VIKING XVII. Oslo.

Humbla, P. & Thomasson, H., 1934. Äskekärrsbåten. Göteborgs och Bohusläns Fornminnesförenings tidsskrift 1934. Gotenburg.

Haasum, S., 1974. Vikingatidens segling och navigation. Stockholm.

Kielland, K., 1954. Hvordan seilte vikingene med sine båter? VIKING XVIII Oslo.

Lindqvist, S., 1941-2. Gotlands Bildsteine I-II. Stockholm.

Nicolaysen, N., 1882. Langskibet fra Gokstad ved Sandefjord. Oslo.

Olsen, O, & Crumlin-Pedersen, O., 1967. The Skuldelev Ships II: Acta Archaeologica XXXVIII Copenhagen.

Shetelig, H., 1917. Skibet. In Osebergfunnet I. Oslo.

Shetelig, H., 1917. Tuneskibet. Oslo.

Thorvildsen, K., 1957. Ladby-skibet. Copenhagen.

KEYWORDS

Medieval; Viking Age

Northern Europe

Viking ships; rigging; sails

Excavated evidence; iconography; documentary sources; experimental archaeology.

12. LE NAVIRE MEDIEVAL A TRAVERS LES MINIATURES DES MANUSCRITS FRANCAIS

Christiane Villain-Gandossi

Résumé

Les miniatures des manuscrits à peinture examinés proviennent des fonds de la Bibliothèque nationale de Paris et de la Bibliothèque de l'Arsenal. Les limites chronologiques de l'enquête sont : le IXe-XVe siècles. D'après ces représentations, la classification des bâtiments (simple, si on l'oppose à la complexité de la terminologie des textes écrits) se résume ainsi : simple nacelle, embarcation avec un mât, bâtiment doté de superstructures, type de la nef-carraque, type de bâtiment à fond plat, bâtiment côtier et de rivière, bâtiment long, bâtiment idéalisé ou imaginaire... Ces représentations permettent de préciser l'évolution des types, au niveau de la coque, de l'appareil de gouverne, des superstructures, du gréement, des apparaux.

Pour l'étude de ce champ particulier de l'iconographie : celui des représentations du navire médiéval, nous avons choisi un certain nombre d'enluminures contenues dans des manuscrits à peinture et appartenant au fonds grec, latin et français de la Bibliothèque nationale de Paris et à celui de la Bibliothèque de l'Arsenal.

Le terminus *a quo* de l'enquête est le IXe siècle : nous le savons déjà: à travers les représentations de cette époque, nous n'aurons guère de témoignages concrets de la vie matérielle des temps carolingiens. Il s'agit plutôt d'une survie de l'iconographie de la Basse Antiquité avec ça et là de discrètes interventions des artistes de cette époque et au-delà des témoignages antiques, l'insinuante présence patente ou clandestine de Byzance. Le terminus _post quem_, c'est la fin du XVe siècle, l'époque des mers fermées, avant l'interpénétration des mondes océaniques.

La classification par type de bâtiment - à travers les représentations des miniatures - nous est apparue peu aisée à établir. W. Frahm (_Das Meer und die Seefahrt in der altfranzösischen Literatur_, Götingen, 1914) avait recensé, par exemple, à travers les sources littéraires françaises, jusqu'au XVe siècle, tous les noms des types de bâtiment qui apparaissent chez les chroniqueurs. On retrouve chez ceux-ci le souci d'évoquer avec précision, de donner un terme spécifique à tous ces bâtiments. D'après cette nomenclature que nous nous sommes attaché à compléter par ailleurs (Ch. Villain-Gandossi, La mer et la navigation maritime à travers quelques textes de la littérature française du XIIe au XIVe siècle, _Revue d'Histoire Economique et Sociale_, 1969, n° 2, p. 150-192; Navires du Moyen Age, _Archeologia_, 1978, n° 114, p. 8-23), on distinguait très nettement, le long des rivages de l'Europe, deux traditions dans le domaine de la construction navale: celle des mers septentrionales et de la Baltique, celle de la Méditerranée. Malgré les importants emprunts, intervenant entre ces deux traditions à l'époque des Croisades, c'est seulement à la fin du XVe siècle, que leur fusion donna naissance à une autre lignée qui devait les surpasser, celle de la tradition atlantique.

Chez les chroniqueurs, on relève donc, avant le XVe siècle, pour la tradition des Mers du Nord, les noms des types suivants : cogue, batel ou bateau, eschief, bac, esnèque, kanart, baleinier, hokebot, scute, hulque.... Pour la tradition méditerranéenne, on peut citer : parmi les bâtiments "ronds" , la nef, la caraque, la barge; parmi les bâtiments"longs", la galée, la galiote, la galéasse, le chat, le brigantin, le pamphyle, la sagette, le lin; parmi les types de bâtiments intermédiaires ou indéterminés : le dromon, le chaland ou sélandre, l'huissier, la targe ou tafourée, la taride, la busse ou busse-nave, la barque, le colombet, la barbote, la grip, le scaphe. Il est certain qu'un grand nombre de ces types devaient présenter des caractéristiques similaires. On est frappé, quand on lit ces textes, par le nombre élevé de synonymes, de redondances. En fait, le hiatus

Fig. 12.1 Bib. nat. Paris, ms. grec 510, f°3 (IXe siècle).
(Bibliothèque Nationale)

Fig. 12.2 Bibl. nat. Paris, ms. fr. 301, f° 153 v° (XIVe siècle) "Les Livres des histoires du commencement du monde" (Bibliothèque Nationale)

Fig. 12.3 Bibl. nat. Paris, ms. fr. 103, f° 1 (XVe s.) "Le Rommant de Tristan et Yseult" (Bibliothèque Nationale)

est profond entre la terminologie relevée dans ces textes établis par des hommes cultivés et les représentations le plus souvent stéréotypées d'artistes moins au fait des choses de la mer.

Dans un premier temps, nous allons procéder à une classification schématique des types de bâtiments, selon les "silhouettes" qu'en donnent les représentations; en second lieu, nous essaierons de préciser l'évolution de ces types, au niveau de la coque, de l'appareil de gouverne, des superstructures, du gréement, des apparaux.

I. Classification des représentations de bâtiments

1) simple nacelle

L'expression de l'esquif le plus rudimentaire est donnée surtout dans les manuscrits carolingiens. La coque est en forme de croissant, avec le souci toutefois de distinguer la proue de la poupe. Le bordage est à clin. L'appareil de gouverne - disposé généralement à tribord - est constitué par un simple aviron, tenu comme une pagaie. On note quelquefois la présence d'une petite barre transversale à l'extrémité supérieure de la hampe.

2) embarcation avec un mât

Cette simple nacelle peut être dotée d'un mât, souvent à crosse, et plus ou moins déporté hors de l'axe médian du bâtiment. Ces représentations schématisées sont dotées, soit d'une voile carrée, soit d'une voile latine.(Sur le document de la figure 1, du IXe siècle, apparaît l'équivalent de modestes tartanes, non pontées. On y note l'indication de bancs, un mât à crosse, une voile latine et la présence d'étais et de bras).

3) bâtiment doté de superstructures

Sur des coques encore archaïques, des pateformes plus ou moins élémentaires sont intégrées (Noter sur la figure 2 ces représentations sytématiques; sur la figure 3, les plateformes de combat sont intégrées; noter l'indication soignée du bordage à clin, sans

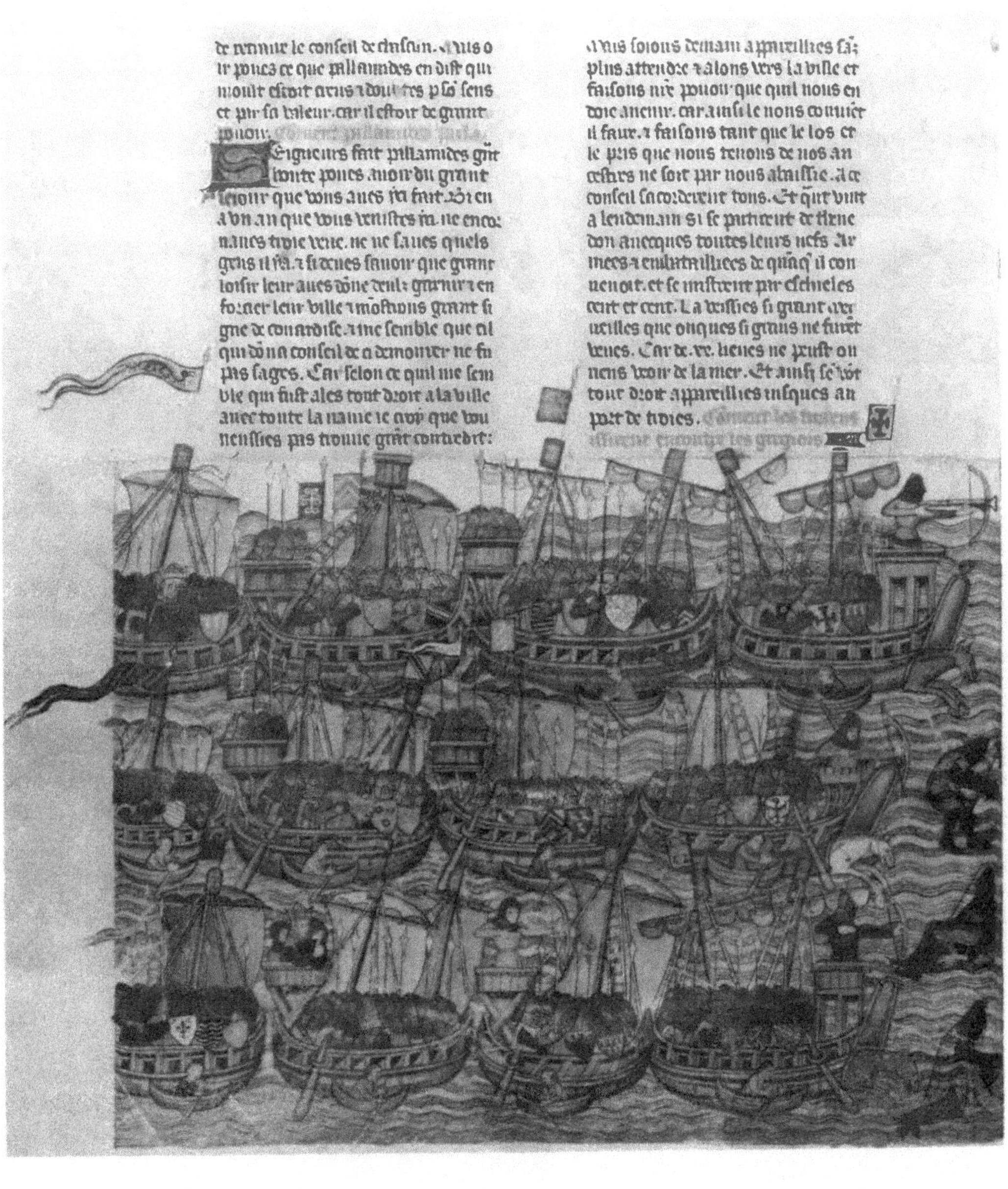

Fig. 12.4 Bibl. nat. Paris, ms. fr. 301, f° 58 v° (Bibliothèque Nationale)

Fig. 12.5 Bibl. nat. Paris, ms. lat. 6142, f° B v° (fin XVe s.)
(Bibliothèque Nationale)

exagération dans l'expression des joints verticaux).

Dans cette rubrique, on peut classer, par exemple l'"huissier" - qui doit son nom à la porte ("l'huis") dont sa coque était percée -, la "targe" et la "tafourée" qui servaient principalement au transport des chevaux. Sur la figure 4, on note que ces bâtiments de service sont ras sur l'eau et conçus spécialement pour des manoeuvres de débarquement, de transport d'hommes et de chevaux. Le sabord de charge est à l'arrière (car, c'est par la poupe que les bâtiments viennent s'échouer). Le gouvernail est latéral. Les triangles sur la coque sont des sortes de fargues. Quant aux plateformes de combat, elles présentent un aspect provisoire : les hommes armés sont là pour protéger la descente.

4) type de la nef-carraque

Dans cette rubrique, nous faisons entrer des types divers. Nous avons mis là, malgré les irrégularités dans les silhouettes, tout ce qui tend à exprimer la volonté d'indiquer un gros bâtiment. Il n'y a guère d'exemples avant le XVe siècle (un exemple du XIIe-XIIIe siècle : la nef génoise sur laquelle s'est embarqué à Constantinople, Conrad, marquis de Montferrat, qui se portait au secours de Tyr, en 1187 : Bibl. nat., ms. lat. 10 136, f° 108).

Au début du XIVe siècle, la nef méditerranéenne se transforme complètement sous l'influence du navire nordique : elle abandonne la voile triangulaire pour la voile carrée, adopte le gouvernail d'étambot. Au grand mât du milieu s'ajoutent les petits mâts d'arrière et d'avant (d'un grand secours pour favoriser, l'un le lof et l'autre, l'abattée). Elle se révélait beaucoup plus aisée à conduire que la nef latine dont les deux grandes antennes exigeaient pour les manoeuvres un équipage important. Sur la figure 5, de la fin du XVe siècle, on est en présence d'une représentation naïve, mais avec le souci du détail: la coque - fantaisiste - montre de grosses préceintes, des défenses, des extrémités de baux (que représente le carré sous l'écubier ?). Dans les superstructures, on note une sorte de galerie la-

Fig. 12.6 Bibl. nat. Paris, ms. fr. 22 547, f° 65 (1470) "Histoires du grant Alexandre de Macedone par Quinte-Curce..." (Bibliothèque Nationale)

térale, de passavant; ces claire-voies servent à abriter les hommes au combat. Le gréement est désormais classique. Le grand mât principal, traité ici comme un tronc d'arbre, est hors de proportion. On y voit de grosses roustures, une vague indication de coins de pied de mât et une boule de racage à mi-mât . Le mât de misaine est miniaturisé; le mât d'artimon a un gréement latin; une sorte de beaupré, à l'avant, sert à fixer l'étai du grand mât. La vergue du grand mât est en deux éléments. Les haubans à colonne sont fixés extérieurement. On note la présence de bras, de doubles balancines. La hune est demesurée; le filet de protection sert contre les traits adverses. A l'avant du bâtiment, le câble-chaîne avec grappin désigne-t-il une sorte d'ancre de détroit ou sert-il pour l'abordage ? Notons aussi la présence de câblesen chanvre sortant par les écubiers et d'une échelle hors le bord.

5) type de bâtiment à fond plat

Il sert pour le transport, les manoeuvres de débarquement, le combat. Ce type de bâtiment à fond plat, pouvant s'échouer, très large pour contenir beaucoup d'hommes, est représenté sur la figure 6:

Sur la coque, présentant une certaine symétrie, sont dessinées des préceintes : il s'agit donc d'un bordage à franc-bord. A l'intérieur, on voit les membrures (varangues et allonges). Dans la deuxième embarcation, les rameurs sont au-dessous du plat-bord; ils doivent ramer debout. Notons aussi la petite figure de proue, l'escalier pour accéder au plat-bord, la passerelle pour aller à terre. Le mât et sa petite voile - qui n'est pas là pour augmenter le sillage, mais comme voile de manoeuvre - sont uniques. Les étais sont représentés.

6) bâtiment long

Les représentations de bâtiments de types allongés sont rares et schématiques.

Le bâtiment mixte, très antiquisant, représenté sur la figure 7, provient du manuscrit de la Bible de Charles le Chauve. Il res-

Fig. 12.7 Bibl. nat. Paris, ms. lat. 1, f° 3 v° (IXe s.) "Bible de Charles le Chauve" (Bibliothèque Nationale)

Fig. 12.8 Bibl. nat. Paris, ms. fr. 5594, f° 267 v° (XVe s.) "Passages faiz oultre mer par les Francois contre les Turcqs et autres Sarrazins et Mores oultre marins" (Bibliothèque Nationale)

Fig. 12.9 Bibl. nat. Paris, ms. lat. 7239, f° 61 v° (XVe s.) "...tractatus de re militari et de machinis bellicis" (Bibliothèque Nationale)

semble à la liburne : il traduit des influences byzantines. On y note la volonté d'exprimer tout un ensemble de rames; elles sont établies sous une sorte de fargue et passent dans de petits sabords. Le motif de cette fargue qui relève le niveau du plat-bord est repris vers l'arrière, sans doute, par souci d'esthetisme. Le gouvernail latéral double sort par une jaumière (le timonier est à l'abri sous une sorte de teugue): il est manoeuvré de l'intérieur. Le gréement est carré. Le mât est important, car il est disposé de plusieurs éléments d'après les roustures. La vergue est en bois tourné. La voile est composée de bandes de toile cousues en damiers, à joints rompus, typiques de la Méditerranée. Le document ne présente ni ris, ni étais, mais des bras et des balancines qui reposent sur plusieurs points de la vergue. Les trois cordages sur bâbord représentent-ils une sorte de haubans ?

7) <u>bâtiment côtier et de rivière</u>

Sa représentation n' a guère attiré les enlumineurs. Sur la <u>figure 8</u>, le bachot fluvial, ou bac, possède un avant relevé et plat pour accoster sur la berge. Le gouvernail axial est posé sur une espèce de chevalet.

8) <u>bâtiment idéalisé ou imaginaire</u>

Sous cette rubrique, on peut faire entrer tous les bateaux de rêve, comme les arches de Noé, les bâtiments de parades, les bâtiments imaginaires (voir par exemple, <u>figure 9</u> , une représentation de Mariano Taccola). Par ailleurs, certains documents témoignent d'une ignorance absolue du sujet : on est en présence alors de véritables "nefs des fous"...

Ce classement est sommaire; il doit pouvoir être affiné, lorsqu'on aura recueilli le plus grand nombre d'informations écrites pour chacun des types et après une confrontation attentive de l'image et des textes. L'examen de ces miniatures permet de cerner l'évolution des types, évolution qui va se marquer dans la complexité des formes à plusieurs niveaux.

II. Evolution des types

1) au niveau de la coque

Les représentations permettent de déterminer la nature du bordé : clin, franc-bord, bordages agrafés. D'une manière générale, il y a présomption de clin, lorsque celui-ci est suggéré par un trait d'ombre appuyé, par une seule ligne de clous ou lorsque la miniature représente un bâtiment avec des extrémités très relevées : c'est alors la seule possibilité de mode d'assemblage. Lorsqu'on note des préceintes, il y a présomption, en revanche, de franc-bord.

Sur la coque, on va relever divers "accidents" : des extrémités de baux, des porte-haubans, des tablettes, des écubiers, des passages d'avirons ou sabords de nage, des trous de jaumière, les défenses (dans la partie centrale, sur l'arrière ou sur l'avant : elles permettent de juger de l'importance du bâtiment) et, bien entendu, les sabords d'artillerie. Mais en général, jusqu'au début du XVIe siècle, l'artillerie est dans les châteaux, non pas seulement, parce qu' on ne sait pas percer la coque, mais aussi parce que l'on est encore dominé par l'instinct militaire du Moyen Age : surplomber l'adversaire. Les châteaux avant et arrière jouent en quelque sorte le rôle de petits châteaux forts. Mais après la substitution du cuivre, puis du bronze au fer forgé qui permettent d'accroître la puissance des pièces, c'est l'usage de l'artillerie qui entraîne la disparition de la nef ou son évolution vers d'autres types, avec l'élévation de la partie du milieu.

On peut apprécier également la nature de la tonture : tous ces bâtiments sont destinés à être traînés à terre; il faut donc être prudent, lorsque la tonture est exagérée. Le rapport longueur-largeur est difficile à apprécier , car les vues sont le plus souvent de profil.

2) au niveau de l'appareil de gouverne

Les miniatures vont permettre de confirmer les distinctions à établir entre les divers systèmes. L'aviron de gouverne - qui est amarré généralement au plat-bord par une estrope - peut être manié comme une simple godille pour les besoins d'une évolution; il se

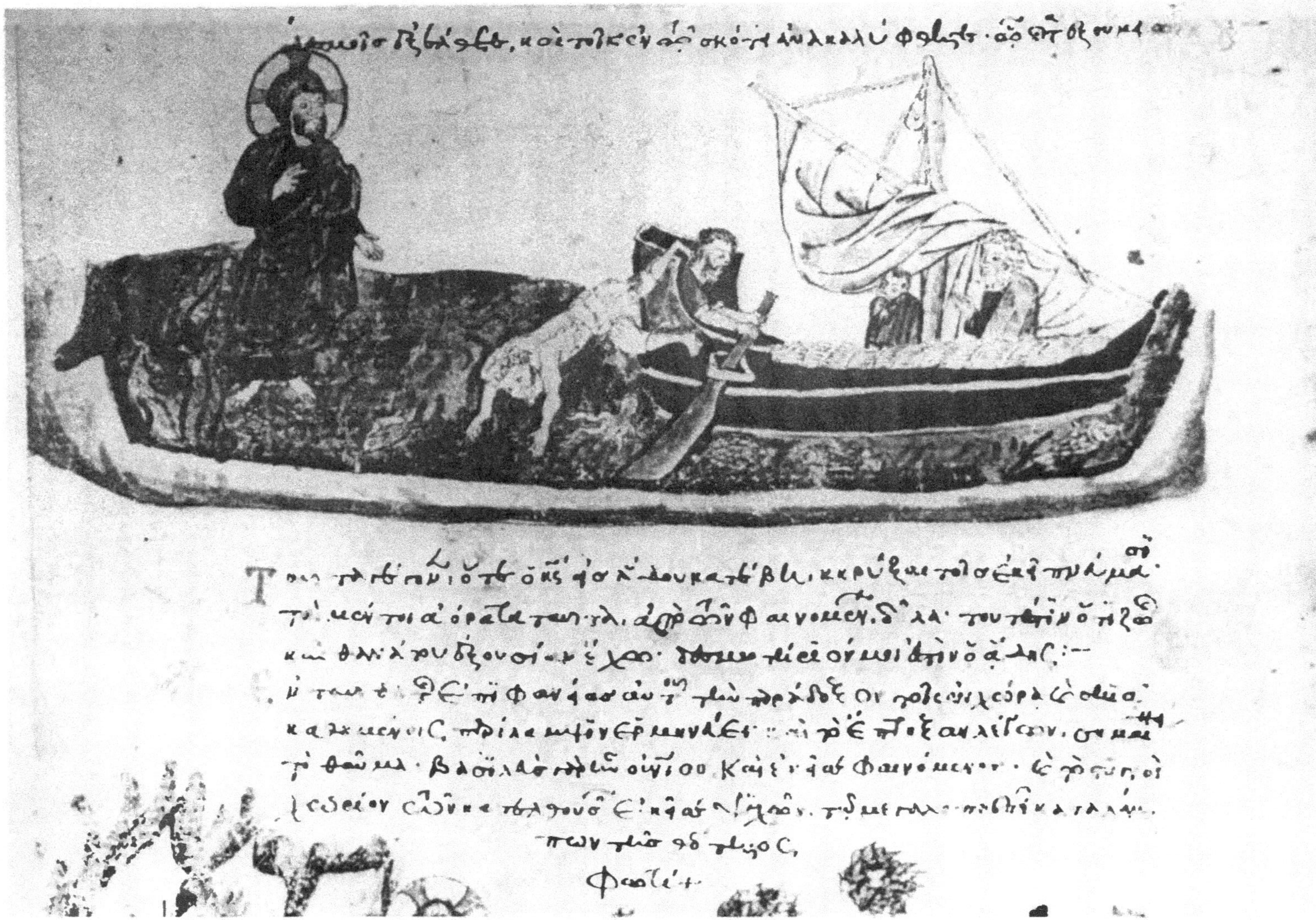

Fig. 12.10 Bibl. nat. Paris, ms. grec 135, f° 220 v° (1368)
(Bibliothèque Nationale)

meut dans toutes les directions. La rame-gouvernail ou gouvernail-rame ou gouvernail latéral est en forme d'aviron à large pale; elle est fixée en général, à tribord arrière, près de l'étambot, par des filins de retenue et de balancement et munie d'une barre perpendiculaire à sa hampe qui lui permettait de faire varier l'angle du plan de la pale. La fixation sur la coque peut être assurée par un arceau de fer et par un trou percé dans la timonière - qui désigne la pièce de bois mobile autour de son axe horizontal. Cette rame-gouvernail ne fonctionne nullement comme un aviron, mais bien comme un safran, du fait de son obliquité plus ou moins grande, par rapport aux filets liquides.(Voir un exemple de timonière, figure 10).Quant au gouvernail axial d'étambot, il a vivement frappé les artistes et est devenu un thème iconographique en soi, avec toutes les exagérations attendues (Voir figure 11 le gouvernail (amovible, peut-être) de cette miniature du XVe siècle).

La miniature de 1307 (figure 12) met en évidence la forme de la barre de gouvernail passant par la jaumière (noter la présence de l'habitacle sur le bâtiment de gauche).

3) au niveau des superstructures

La fonction première des superstructures apparaît d'une manière évidente à travers ces représentations : il y a eu évolution, lorsqu'on a envisagé de se battre sur l'eau (il faut noter une évolution parallèle du développement de la hune). La première étape s'est réalisée, lorsque la coque s'est prolongée pour former le gaillard d'arrière. Avec le bordage à clin, on aura toujours une plate-forme en surplomb;avec le bordage à franc-bord, on aura un élargissement des formes de l'arrière. En tout cas, les châteaux constituent un thème iconographique en eux-mêmes. Il serait intéressant de définir et de classer leur décor : les ajours, les ogives, les treillis, les créneaux... Leur aspect, sur les miniatures, est généralement, plus plaisant que militaire.

A l'avant, la plate-forme de combat est terminée par un triangle, à l'image du bec de l'Antiquité; d'où un aspect plus rapporté que pour le château arrière. Souvent cette plate-forme ne s'incorpore en rien à la coque proprement dite. Cet éperon est plat par en-dessous. Un élément en fer-à-cheval, sorte d'ouvrage léger, termine

Fig. 12.11 Bibl. nat. Paris, fr. 77, f° 386 v° (XVe s.) Chroniques d' Angleterre de Jehan de Wawrin (Bibliothèque Nationale)

Fig. 12.12 Bibl. nat. Paris, ms. fr. 2810, f° 263 (1307) Livres des Merveilles (cliche Musé Marine Paris)

Fig. 12.13 Bibl. nat. Paris, ms. fr. 5594, f° 112 (Bibliothèque Nationale)

Fig. 12.14 Bibl. nat. Paris, ms. fr. 365, f° 89 (XVe s.) (Bibliothèque Nationale)

le gaillard d'avant. On y accède avec une échelle.

La partie centrale, c'est l'artillerie qui va la revaloriser. On aura la tentation d'y mettre les sabords, car ainsi l'artillerie sera plus maniable (il faut d'ailleurs placer assez bas les poids lourds des canons en fonte). Les châteaux , devenus superflus, disparaissent, tandis que par compensation, la muraille de la partie centrale se relève.

Le gaillard d'arrière est davantage intégré à la coque. Sur la figure 13, on note la manière dont se fait le raccordement du château arrière avec le pont proprement dit, avec un élément en fer-à- cheval. Quant à l'avant, on remarque sur la tête de l'étrave les deux pièces de bois transversales qui déterminent une surface en triangle avec la pièce de bois longitudinale. Il faut des bordages pour remplir ce triangle, d'où le remplissage de cette surface exécuté en bordages à clin, bouchant le vide.

4) au niveau du gréement

La voile vraiment motrice est celle du grand mât. Celui-ci est toujours représenté d'une manière exagérée dans les représentations. Cette grand'voile - et les miniatures en témoignent- est perfectionnée par une petite voile placée sur l'avant, dont l'utilisation est délicate, à cause du déséquilibre produit. La voile qui s'établit sur l'arrière joue sur le centre de gravité. L'aboutissement, ce sont les trois mâts verticaux; le mât de beaupré sert d'appui à l'étai du mât de misaine (pour éviter le tangage). Ce n'est qu'au XVIe siècle qu'y apparaîtra la civadière. Les miniatures mettent en évidence la façon dont les voiles sont enverguées (à la manière d'anneaux de rideaux ou avec des"yeux de pie"), la façon dont elles sont assemblées... On voit les ris, pris en partie basse ou par souci d'esthetisme, sur toute la surface . (Voir sur la figure 3, au premier plan, la bonnette boutonnée trop haut).

Pour les mâts, on dispose de certains renseignements : le nombre de mâts, les élévations sur un même mât, les mâtures d'une seule venue ou non, la présence de roustures... Sur la figure 14, du XVe siècle, on distingue le petit mât de hune avec sa voile.

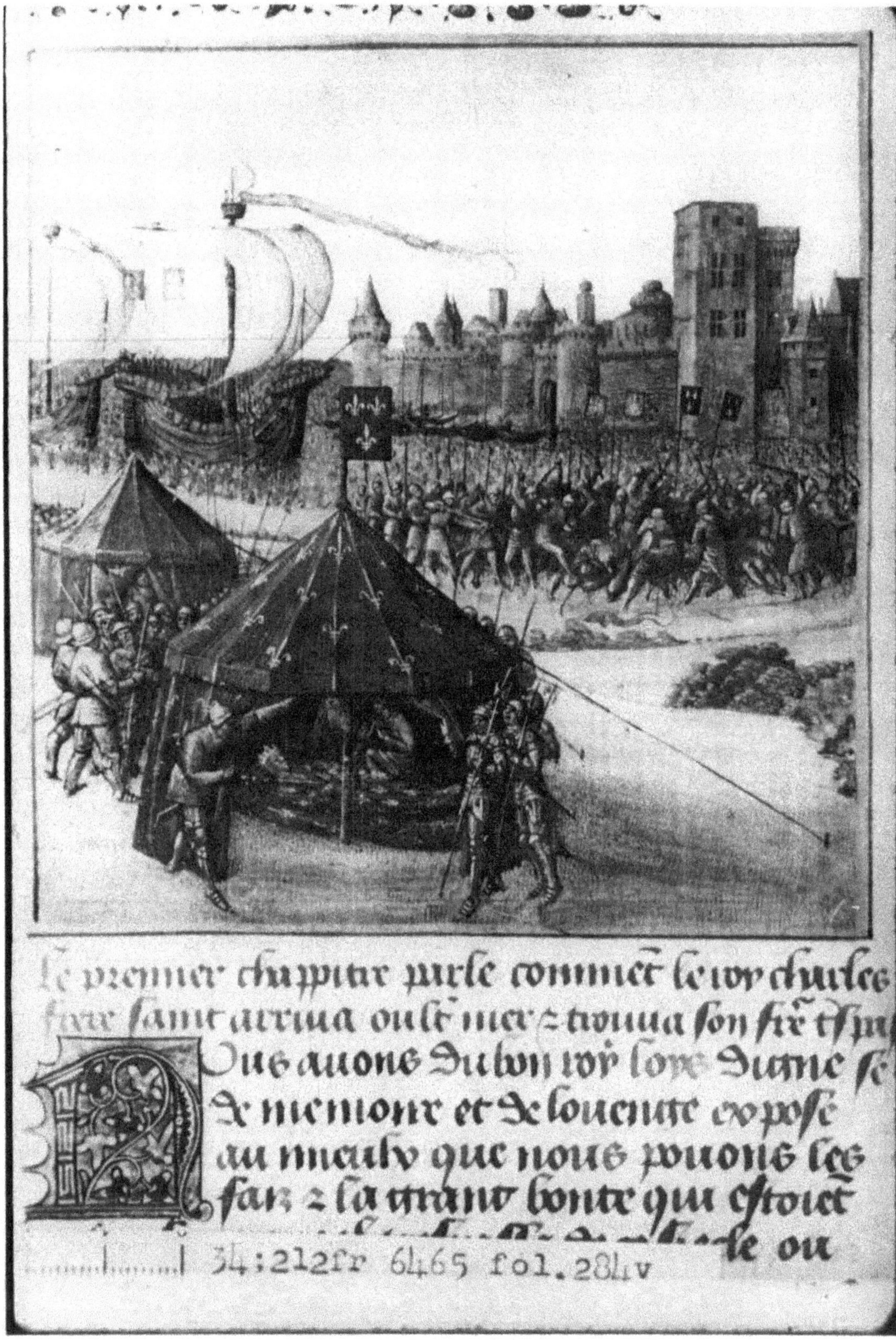

Fig. 12.15 Bibl. nat. Paris, ms. fr. 6465 284 v° (XVe s.) Chroniques de Saint Denis (Bibliothèque Nationale)

Les vergues seront souvent en deux éléments juxtaposés formant saillie; ce n'est que plus tard qu'elles apparaîtront taillées d'un seul tenant.

Quant aux manoeuvres, leur pérennité d'emploi est évidente. Mais les peintres représentent rarement la totalité des filins; ils font un choix. Ce sont les bras, les étais, les balancines, les écoutes qui sont le plus souvent représentés . Ils omettent souvent les cargues : les filins lâches nuiraient à l'ensemble de la composition. Les haubans seront des haubans volants avant d'être des haubans fixes (mis à demeure avec les caps de mouton). En tout cas, ris et haubans donnent une notion des dimensions des bâtiments. Il est difficile de voir l'aboutissement des manoeuvres courantes, car rares sont les vues par " en-dessus" qui seraient si utiles pour révéler l'aménagement du pont. Sur le document de la <u>figure 15</u>, on note des détails intéressants sur le gréement : une écoute sur la voile dans l'axe, des balancines doubles, des bras, de grandes cargues, des haubans (fixés à l'intérieur ou à l'extérieur, c'est difficile à dire, car leur point de fixation est masqué par des pavois), des étais sur l'arrière (noter également le hunier miniaturisé).

5) <u>au niveau des apparaux</u>

Les agrès et apparaux n'apparaissent le plus souvent qu'à l'état d'indice, sauf peut-être pour le détail des ancres.

Voir sur le document de la <u>figure 16</u> - qui symbolise la navigation en direction- une représentation de compas.

Après l'examen d'un certain nombre de miniatures, il est permis de dégager quelques caractères spécifiques de leurs représentations :

<u>influence du champ :</u> C'est une règle prudente dans l'interprétation des documents schématisés de penser à ce que pouvaient être d'une part, les limites dans lesquelles devaient s'inscrire le modèle, d'autre part, les espaces qu'on y devait peut-être ménager. En général les bateaux des miniatures sont plus ramassés que dans la réalité. La

Fig. 12.16 Bibl. nat. Paris, ms. fr. 2810, f° 188 v° (Bibliotheque Nationale)

Fig. 12.17 Bibl. nat. Paris, ms. fr. 5594, f° 34 v° (Bibliothèque Nationale)

Pou de temps apres
ce que le conte de
neuers fut reuenu
en france fut
enuoie en ambaxade vng
seigneur grec appelle Catotu
seno de par lempereur de con
stantinople que on nommoit
Carmanoli. Le quel supplioit
au roy charles quil voulsist
pour honneur de la foy xpienne
et amour de nre seigneur Ihu
crist lui donner secours et aide
contre les turcqz ou que autre
ment toute lempire de grece
seroit bref perdue par ce quil
ne leur pouoit plus resister.
Et pour celle chose mesmes
enuoyerent aussi leurs deuers
le roy Les veniciens et Jenne
uois remonstrans le grant
peril ou estoient tous les xpien

Fig. 12.18 Bib. nat. Paris, ms. fr. 5594, f° 269 v° (Bibliothèque Nationale)

tendance à exagérer la hauteur aux dépens de la longueur est de règle, en pareil cas. Souvent, consciemment ou non, l'artiste subit l'influence du mouvement général de la composition; la pression du cadre ou des bords se fait sentir à distance, sur des motifs qui sont à l'aise loin des marges. Ceci se remarque notamment, lorsqu'il s'agit de décorer une bande horizontale plus ou moins allongée; les bateaux, au lieu de se cambrer, prennent alors une allure de longs fuseaux sans grande tonture.

symétrisation : par exemple, les bâtiments n'ont ni avant, ni arrière : on s'éloigne délibérement des réalités (voir figure 17).

stylisation, schématisation : le document de la figure 18 est un bon document et pourtant, l'exécution semble être faite en pierre; les défenses sont vraiment exagérées: elles sont là comme élément décoratif; les mâts sont d'une hauteur trop importante : d'un bon tiers à l'oeil; les voiles sont disposées trop haut, sans doute pour que ne soient pas masqués les détails de la coque; les représentations de manoeuvres sont stéréotypées, conventionnelles.

prolifération : certains éléments attirent l'oeil de l'artiste; ces éléments, il va les multiplier : ce sont les clous du bordage à clin, utilisés en motifs décoratifs, ce sont les garcettes de ris qu'on multiplie sur toute la surface de la voile; ce sont les roustures des mâts appliquées systématiquement; ce sont les écubiers qui prolifèrent...

confusion, ignorance : l'image représentée sur la figure 7 est une copie très déformée de quelque galère romaine. Dans les documents qui ont pu servir de modèle à l'origine, l'apostis, l'encorbellement prismatique qui porte les rames, est représenté en perspective. Dans la miniature, cette face arrière a été disjointe du reste; elle est devenue une espèce d'écubier en losange, tout à fait invraisemblable. En effet, le gouvernail, sortant ainsi à peu près perpendiculairement à la surface de la coque, ne serait pas manoeuvrable. Qu'un copiste suivant vienne arrondir le contour en losange, il ne restera plus de trace de la provenance de celui-ci et le document n'offrira plus au chercheur qu'une indication fausse, quant à la disposition du gouvernail.

Il est évident que l'on pourra noter une évolution à travers ces représentations, depuis les enluminures pré-romanes. A partir de 1350 environ, on assiste à une conquête de la réalité, des apparences sensibles; on trouvera le souci de la part du peintre de représenter les spectacles qu'il a sous les yeux. Chez les enlumineurs gothiques, les formes s'assouplissent, les silhouettes plates disparaissent, les proportions et les perspectives tendent à exprimer non plus l'idée que l'on se fait des objets, mais l'aspect sous lequel on les voit.

La comparaison avec les autres apports de l'iconographie médiévale : vitraux, peintures, gravures, monnaies, sceaux, graffiti, bas-reliefs... ne sera pas à l'avantage des enluminures, notamment pour les sceaux, pour les graffiti, mais surtout pour les peintures, les gravures : que l'on songe à l'admirable gravure sur bois de Jacopo Barbari (Museo storico navale de Venise) ou au cycle de Sainte Ursule de Carpaccio ou encore à la vue en perspective de Boticelli de la chapelle Sixtine...

Ne serait-il pas souhaitable que soit mise à l'étude une grille d'analyse systématique, valable pour tous les types de sources qui permettrait de comparer les documents selon des critères cohérents ?

MEDIEVAL SHIPS AS SHOWN BY ILLUMINATIONS IN FRENCH MANUSCRIPTS

Christine Villain-Gandossi

English Summary

This paper presents illuminations from various medieval painted manuscripts located in Paris at the "Biblithèque Nationale" and the "Bibliothèque de l'Arsenal". The centuries covered are from the 9th to the 15th. But the illustrations will by no means be chronologically classified.

1. The first classification is based on the types of boats represented. Unfortunately, such a classification must be made without being able to match the drawings and paintings with the many boat names found in the written evidence of the same period. It seems that the writers were much more knowledgeable than the artists who produced the illustrations.

   - simple open boats
   - boats with one mast and square or lateen sail
   - ships with more-or-less elaborate superstructures
   - type of nef, carrack, cog, not always easy to recognise
   - flat bottom ships
   - coastal craft, river boats
   - long vessels
   - idealized, imaginary boats and ships

2. A second kind of classification allows one to follow the development of various features taken individually:

   - hull shapes and details
   - steering apparatus
   - superstructures
   - rigging
   - gear and equipment

3. Finally, certain remarks are made on specific details of the manuscript illustrations, affected by

   - size
   - habits and traditions of illustrations
   - sketching methods

KEYWORDS

Medieval.
France.
Boats and ships; illuminated manuscripts.
Iconography; shipbuilding; hull; steering; superstructures; rigging.

13. THE USE OF ICONOGRAPHIC MATERIAL IN MEDIEVAL SHIP ARCHAEOLOGY

A. W. Farrell

Summary

This paper seeks to examine the use of secondary, and especially iconographic, material in ship archaeology. The value and limitations of this material is discussed and special attention is drawn to the need to record and interpret this material with as much care as possible.

TYPES OF MATERIAL

Iconographic material is so often referred to by ship-archaeologists, that its value as source material needs no introduction. Frequently, iconographic material is quoted to supplement or elucidate information obtained from examining ancient ship, or boat, finds: examples are almost too numerous to quote but recent examples are to be found in Bonino (1978), Ellmers (1978), and Fenwick (1978).

In localities where primary source material is scarce or non-existent (e.g. Ireland, (Farrell, forthcoming) and France (Fenwick, 1978:193), iconographic material may be the only source available for research.

Because of the partial survival of evidence, iconographic material is an essential source, even in those localities where there is an abundance of primary material. Very little is known about medieval rigging and sail plans for instance, and what information there is available has almost wholly been gleaned from secondary source materials (Humphreys, 1978:78).

Iconographic material is generally not costly to research; one experienced fieldworker can do a great deal of useful work simply by examining, recording and publishing the material within a particular locality, or within one particular medium. There are, however, some very real problems to be faced when working with this material. Only rarely is the material unequivocal in meaning, and interpretation can pose serious difficulties (Humphreys, 1978; McGrail and Farrell, 1979). In the past, too many commentators have glossed over these problems and consequently misunderstood or misinterpreted the evidence. Dating is often a major problem with this type of material. The context of a ship representation, the medium in which it is produced and the reason for its production, are vital factors to consider when iconographic material is discussed.

For ease of discussion, and convenience, iconographic material may be grouped according to the medium employed for depiction. The following fourteen groups are suggested as a basis for organising this material:

1. Stone-carving
2. Graffiti (in stone, wood and plaster)
3. Models
4. Manuscript illumination
5. Wood carving
6. Tapestries
7. Mosaics
8. Stained-glass
9. Coins
10. Seals
11. Tokens
12. Murals
13. Metalwork
14. Portable objects

Clearly, this method of classification is not ideal; others may like to divide the groups further, add new groups or merge others (e.g. coins and seals). Although of limited value to the ship-archaeologist, burials may be considered as another group worthy of inclusion. The custom of burying people in graves surrounded by stones set in the shape of a boat is well documented in Scandinavia and appears to have a long ancestry there. Examples are known outside Scandinavia and the subject has been much discussed. Actual boat burials are better classified as a primary source material as many of the boats buried have been extremely well preserved (Gokstad and Oseberg being the obvious examples).

The above classification system is useful in that each group tends to be characterised by certain advantages and disadvantages as regards the survival of the evidence, methods of recording, dating and interpretation. Some of these characteristics may be shared by other groups and they may vary within a group. It is important to realise, therefore, that each ship-representation is unique and must be considered independently.

STONE-CARVING

A wide variety of material is covered by this group; it is varied not only by the material within which it is carved but also by the technique of execution, function, location, skill of portrayal and by the degree and accuracy of detail shown. Some of the difficulties presented by material within this group can be illustrated by the "Bantry boat". This carving is to be found on a carved pillar at Bantry in the S. W. of Ireland (for a more detailed discussion see Johnstone, 1964). Although it is badly weathered and difficult to interpret, the carving is important because of the general lack of material available in Ireland. Despite its worn condition, commentators have been keen to ascribe some form of absolute interpretation to the carving. It can be argued, however, that the nature of the carving and its state of preservation do not realistically allow for positive interpretation.

Fig. 13.1 The Bantry Pillar, Co. Cork, Ireland

Fig. 13.2 Boat carved on 'Prior's Door', Ely cathedral. (Courtesy of Mr. Holmes, Ely)

Commentators (Ellmers, 1969:A6617g, Greenhill, 1976:Fig.75) have sometimes failed to observe the sixth figure in the bow of the boat, even though it was first noted by Johnstone (1964:280) and is still just about visible (Fig. 13.1). Since Johnstone (1964), archaeologists have interpreted the carving as depicting a skin boat similar to the currachs used in Kerry (Ellmers, 1964: 107-8, Greenhill 1976:122, Bowen, 1972:38). Whilst not rejecting this interpretation, there is little concrete evidence to support it. The shape of the Bantry boat is characterised by its pronounced sheer; the Kerry currach does not exhibit such an overall sheer, however. It has also been said that the Bantry boat is shown riding high on the water (Johnstone, 1964:279), like a modern currach, but if the line representing the water is realistically drawn, why does the steering oar not enter the water? The steering oar is set on the boat's port side, yet it passes in front of the steersman's chest, thus setting him in an almost impossible position.

There is a dynamic and realistic quality about the carving, as Johnstone observed, but it would be wrong to ignore the difficulties of interpretation, and any interpretation must be treated with caution. The artist's keen eye for movement tends to hide certain inaccuracies in the carving.

The 8th century date ascribed to this carving has not been seriously questioned but the dating of all stone-carvings of this period in Ireland has been frequently debated.

The boat carved on the Romanesque 'Prior's Door' in Ely cathedral indicates the diversity one may expect from the depiction of ships in stone work (Fig. 13.2). The carvings around the door have been dated stylistically to c. 1140. Unlike Bantry carving, artistic considerations have resulted in the boat being reduced to hardly more than a design motif. It is quite clear that the boat's form has been strongly arched to fit within the surrounding roundel. The oarsmen are set in unlikely juxtaposition in order to create an impression of balance within the roundel. As this simple little carving shows, context and aesthetics are important considerations in the interpretation of this type of material. The function of the carving is equally important.

One may profitably compare the Copenhagen sarcophagus carving with that from Ely. Both are clearly decorative but the Copenhagen carving is believed to be narrative in concept and this has resulted in an intrinsically realistic and well executed depiction of 3rd century A.D. merchant shipping (for a useful introduction to this much discussed carving see Toby, 1974).

GRAFFITI

The value of graffiti to the ship archaeologist varies greatly with the individual examples. Freed from formal artistic or symbolic restraints, graffiti are sometimes the product of a knowledgeable and skilled hand. Graffiti are often spontaneous and individualised pieces of work standing outside artistic conventions, consequently it is difficult to date them stylistically (as was attempted for the Khirbet Rafi graffiti, Kloner and Wachsmann, 1978:227-9).

The major value of this source material is its general availability and its potential accuracy (accepting that some graffiti may actually be the work of a seaman or a boatbuilder rather than an artist).

Fig. 13.3 Ship graffito from Dunluce Castle, Co. Antrim, Ireland

Fig. 13.4 a & b Graffiti in wood, from Winetavern Street excavations, Dublin (after O' Riordan)

i. Stone

There is almost certainly a great deal of ship graffiti in medieval stonework yet to be recorded. Pritchard's work on medieval graffiti (1967) gives some idea of the amount and variety of material available. A useful starting point for the study of this type of material might well be medieval ecclesiastical buildings.

The example of graffiti in stonework illustrated here (Fig. 13.3) comes from the entrance to the gatehouse in Dunluce Castle, N. Ireland. The ship is lightly incised and neatly executed by a seemingly competent hand and it illustrates some of the advantages and disadvantages of this type of material (for a more detailed discussion, see Farrell, forthcoming).

The ship has parallels with some of the so-called "galley" stone carvings of 16th century date from Western Scotland (see Steer & Bannerman, 1977). The gatehouse construction displays Scottish influence and the political connections between Scotland and N. Antrim at this time were strong. This graffito may well be a fairly accurate representation of a 16th century "galley" but there are major dating difficulties with the Dunluce graffito. It is unlikely the graffito pre-dates the gatehouse. The graffito may be of recent date but there are strong reasons for supposing that it was buried under earth and rubble until the Department of the Environment cleared the area in 1928. Dating is a recurring problem in assessing material of this type.

As well as drawing and photographing, it is useful to make a rubbing of graffiti where conditions permit. The Dunluce graffito also illustrates the necessity for recording the exact context of the depiction if accurate interpretation is to be made possible.

ii. Wood

Owing to the impermanent nature of this medium, ship graffiti in wood is relatively scarce outside Scandinavia. Dating is not necessarily such a problem as it is with graffiti carved in stone. These Dublin examples have been dated to the 11th century by the context of the structure with which they were associated (Fig.13.4a and 13.4b).

Both examples comes from the Winetavern Street excavations and give some indication of how interesting this material may be (O'Riordan, 1971). The more complete of the two graffiti (Fig.13.4a) is clearly more stylised and crudely representational than the other (Fig.13.4b) which, although experimental, shows a furled sail and lowered yard in a more realistic manner. Both these examples do much to emphasise the nature and importance of shipping in Viking Dublin.

iii. Plaster

Graffiti in medieval plasterwork is an important source material and almost certainly a field worthy of further research. The examples shown here reveal the potential value of ship-graffiti in plasterwork and also highlight some of the problems (Fig. 13. 5).

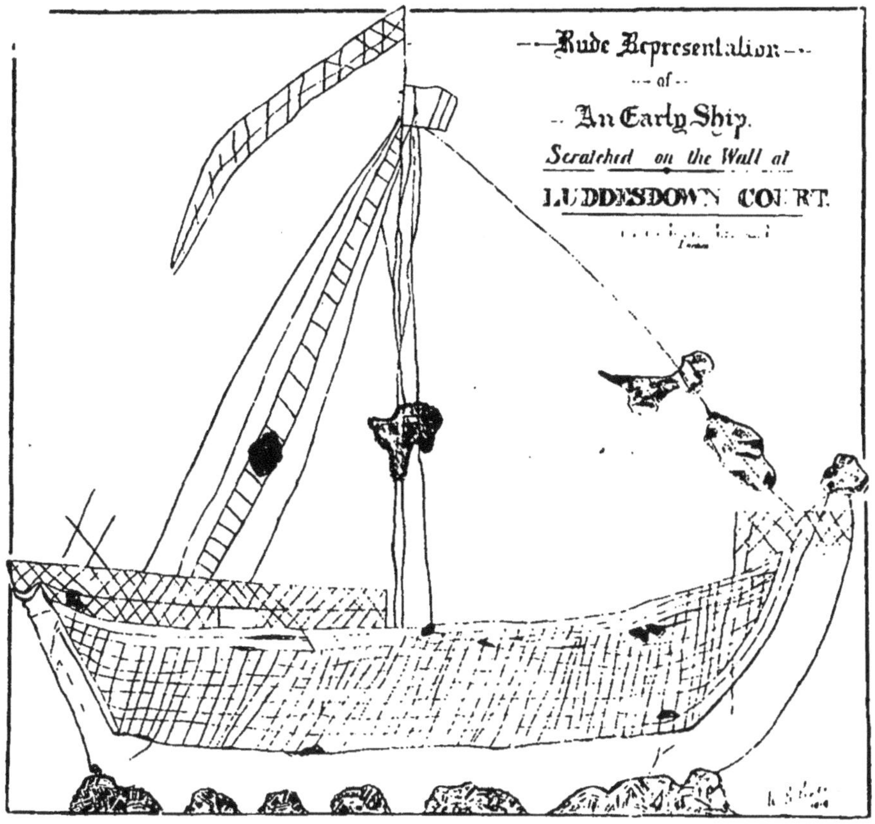

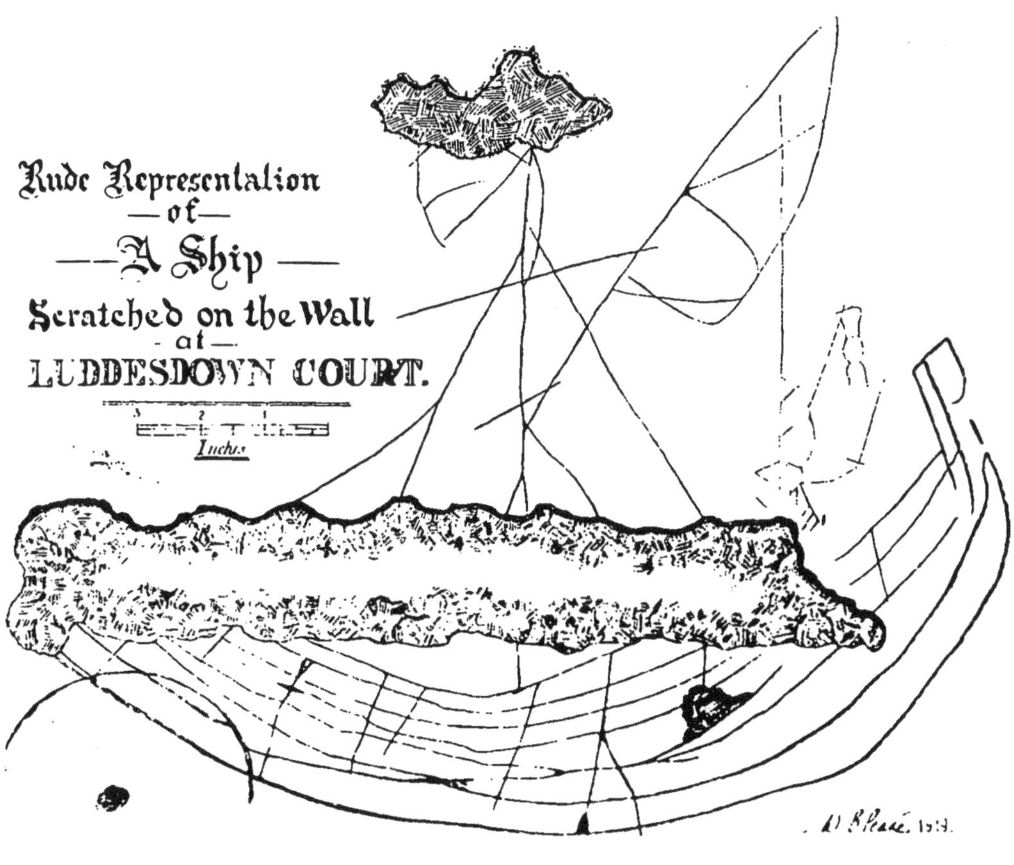

Fig. 13.5 a & b. Graffiti in plasterwork, from Luddesdown Hall, Meopham, Kent (after W. B. Peake)

These two ships are to be found lightly incised in the plasterwork of a great hall of Norman date in Luddesdown Hall, Meopham, near Rochester in Kent. The graffiti are situated on a wall about 1.5 m up from the floor. Although the hall is of Norman workmanship, the plasterwork could be of much later date. The graffiti was found during restoration work, together with other graffiti which has been dated to the 14th century (Peake, 1930). The two vessels shown here would be at home in a 14th century context, but the graffiti itself remains essentially undateable.

The graffiti is of particular interest because it shows two distinctly different ship types which may profitably be compared to the much-quoted 'cogs' and 'hulks' of medieval date. The vessel (Fig.13.5b) with the strongly curving hull shape—reminiscent of the 'hulk' type—is noteworthy because the hull shape is in no sense dictated by the available space, as can be argued for some such representations on seals and coins.

The 'cog'-like vessel (Fig.13.5a) is not only better preserved but also executed with more care, and in considerable detail. The lack of a yard and sail in this context is unlikely to be accidental as everything else about this example indicates it was the work of a keen observer of contemporary ship types.

MODELS

Ship-models are rare in the archaeological record for the prehistoric and medieval periods in Britain and Ireland (they are by no means common in the rest of Northern Europe). Models are to be found in several mediums, including wood, bone and metal, and they have the advantage, for the ship archaeologist, of being three-dimensional.

Models vary greatly in the degree of detail and accuracy they show, and there are often serious difficulties in interpretation, as with the Broighter boat (Fig. 13.6). This model can be dated to the first century B.C. with reasonable confidence and is almost certainly Irish in origin (for a full discussion, see Farrell and Penney, 1975), which thus makes the model especially important. The model would probably have been made by a metalsmith who would not necessarily have been a seaman, Details such as the oars, mast, thwarts and yard are realistic details which any observant craftsman would note, but the gold hull is simply imagined, simply constructed, and lacks detail. The hull has often been interpreted as representing a skin boat (Ellmers, 1969:109-111, Bowen, 1972:38; Greenhill, 1976:121-2) but its simple form defies positive interpretations of this kind. The nature of the find spot—a bog—in this case suggests the model had a votive function and the need for detailed realism would probably not have been a major concern. Again, the need to consider the context of the find and its original function are important aspects of the evaluation process.

Crumlin-Pedersen's comments on the model of a herring 'kagg' in Nibe church are revealing here (1965:89-93). He points out that the model hull was made deliberately narrow to avoid looking clumsy when viewed from below. This model was built in 1703 by a competent craftsman and carefully adapted so that it could be hung in the church at Nibe. Thus seen in its true

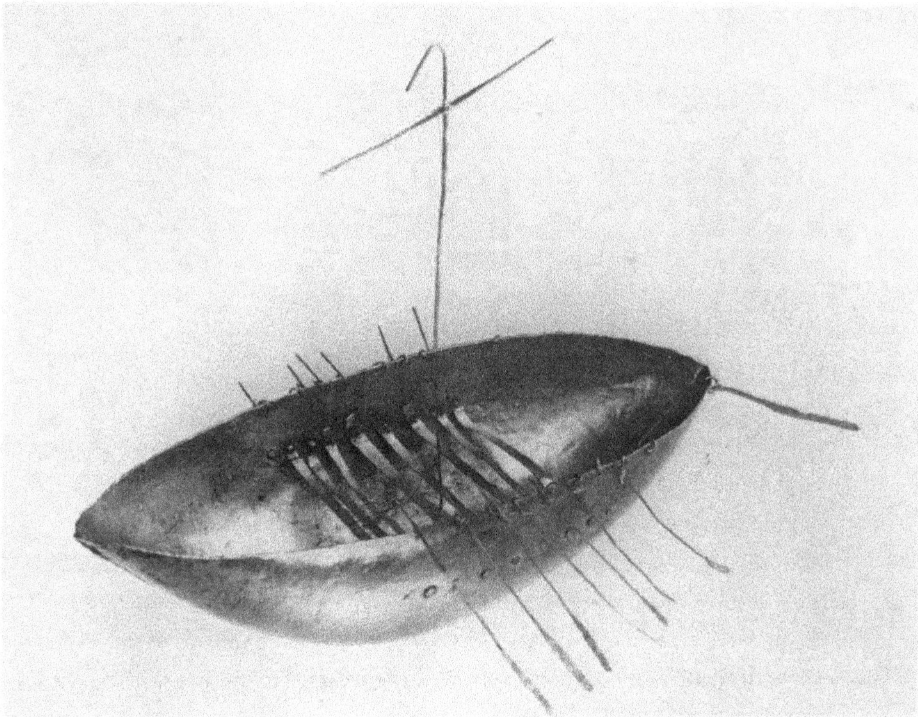

Fig. 13.6 Model boat, from Broighter, Co. Derry, Ireland (National Museum of Ireland)

Fig. 13.7 Illustration from the Cambridge 'Life of Edward the Confessor'

Fig. 13.8 Noah's Ark wood carving, from a miserichord in Ely cathedral

Fig. 13.9 Scene from the Bayeux Tapestry

context, the lack of internal timbers and any curiosities of shape are readily explained. We know little about the original function of many ancient ship models and it is possible that more research in this area would stimulate a revaluation of some models. Any discussion of ship models should take into account the materials employed in their manufacture as these materials may be important factors in deciding the model's final appearance.

MANUSCRIPT ILLUMINATION

Manuscripts are a most lucrative field of study because of the sheer amount of material surviving. It is almost impossible to generalise about the value of this material because the quality, realism and accuracy of the illustrations are so varied. Dating is rarely a serious problem but interpretation is often difficult and the circumstances of production are an important consideration with this material. Very few of these miniatures can have been done by an artist with a trained nautical eye, although many illustrators will have experienced sea-voyages themselves.

The miniatures from a 'Life of Edward the Confessor' (Fig. 13.7) now in the University Library, Cambridge, reveal the value and limitations of this material. As might be expected, the artist appears to have accurately portrayed those features which he understood, or which were visually striking, but shows confusion in more technical areas such as the rigging. The method of attaching the steering oar using a ring (or thong?) is interesting but surely a simplification (see also the illustrations from the Life of St. Albans in Fenwick, 1978:Fig.8.25). The depiction of a 'myke' also helps to interpret the function of similar shaped timbers discovered during the Wood Quay excavations in Dublin in 1975. In this illustration the 'myke' is positioned on top of the sternpost and receives the backstay (or possibly the halliard). The stem ropes, which here seem to function as gammoning for the bowsprit, is a recurring and much debated feature of 13th and 14th century ship representations (see the early volumes of Mariner's Mirror for a full discussion of this feature). The position of the mast and the lacing of the sail to the yard are likely to be accurately portrayed details.

The obvious parallel between this 13th century manuscript and the so-called Paris Apocalypse also raises the problem of copying. Bearing this in mind, it is possible for a ship representation to appear some distance (in time and space) from its original place of inspiration.

WOOD-CARVING

Wood-carving was clearly an important art medium in the medieval period. Many of the problems of interpretation are similar to those described for other sources.

The interesting example shown here (Fig.13.8) depicts 'Noah's Ark' and comes from a miserichord in Ely cathedral. The carving is in such bold relief that it is essentially a three-dimensional model. This particular carving shows how symbolic intent in medieval iconography can displace realism completely. Dated to the 14th century, the castellated towers of the Ark represent the Trinity in Unity, floating in the stormy seas, like God's Church. The hull has here been reduced to a mere platform and is of much less importance than the other pieces of the symbolic jigsaw.

TAPESTRY

The Bayeux Tapestry has been referred to by Frank Stenton as "the simplest and on the whole the most probable version of the story", of events leading up to the Battle of Hastings (Stenton, 1971:577). The Bayeux Tapestry has no rival as a source material in this medium (Fig.13.9) and it has been frequently referred to by ship archaeologists (Olsen & Crumlin-Pedersen, 1967; Brøgger & Shetelig, 1971; Crumlin-Pedersen, 1972; McGrail, 1974; Fenwick, 1978). When considering the ships depicted on the tapestry, it is important to remember the parallels which exist between the designs on the Tapestry and some contemporary English manuscript illuminations (see Stenton, 1957). The possibility of stylisation and conservatism in the depiction of the ships should, therefore, be borne in mind.

MOSAICS

Mosaics are another important source material in ship archaeology and have been much quoted in the past (see for example Casson, 1973; Throckmorton, 1972: Throckmorton, 1973; Bonino, 1978; and Ellmers, 1978). Such mosaics are rare in Britain but this example from Low Ham in Somerset (Fig.13.10) includes a scene depicting the Trojan fleet. The mosaic overlays a cold plunge bath constructed c. 330 A.D.; therefore the mosaic is of later date.

Marsden (1972:132) says of the ships depicted on this mosaic that "Vessels of this kind were guided into the late Roman naval base at Dover ..." but it has been argued that the design was based on an illustration from a contemporary (Mediterranean?) manuscript (Smith, 1970:91). Liversidge (1973:284) has also raised the possibility that the mosaic was laid, or designed by an artist from N. Africa, as it has clear parallels with mosaics from Tripolitania.

The ships are likely to be the products of numerous copyings and could thus be far removed from their original source of inspiration. The appearance of these ships on a mosaic in Roman Britain need not necessarily reflect the type of vessels employed around the coast of Britain in the 4th century. The artistic, social and economic circumstances of mosaic productions may thus limit its usefulness as source material. Uncritical discussion of such material may be misleading in some instances.

STAINED GLASS

As well as the difficulties already mentioned, medieval stained glass windows pose other problems for the researcher. The major problem is one of accessibility and many windows can only be viewed from a distance. Previous descriptions are not always helpful. The value of a close examination should be emphasised because many windows have been subject to repairs in

Fig. 13.10 Mosaic from Low Ham villa

Fig. 13.11 Jonah scene in the east central window in the corona in Canterbury Cathedral

Fig. 13.12 British halfpenny, 1967

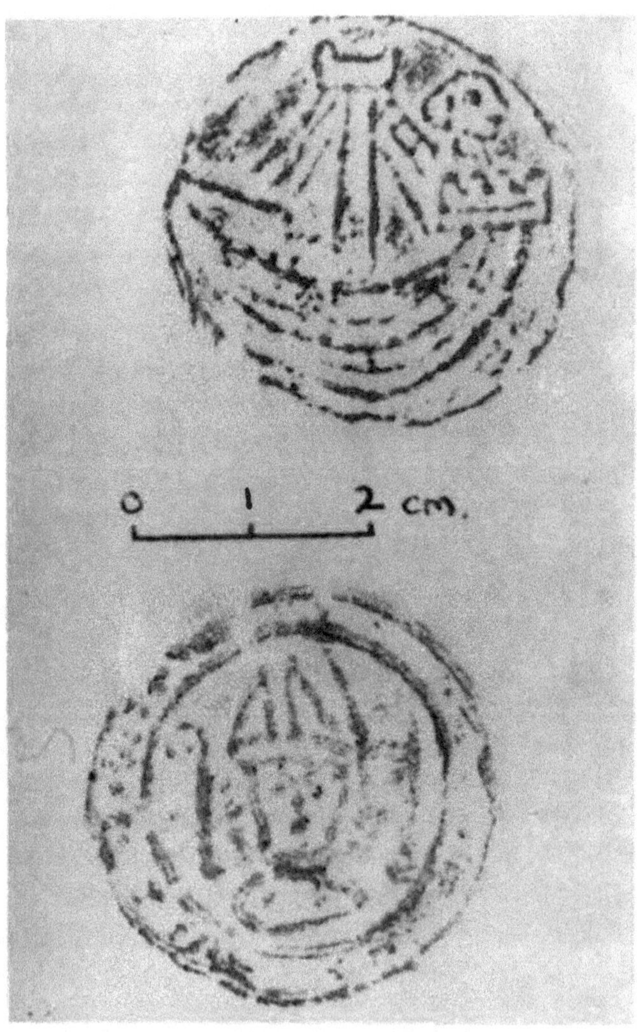

Fig. 13.13 Rubbing of lead " monk's token" now on display in Ely Museum

antiquity and this may influence the appearance of a ship representation. Without close examination, specific features may be ascribed a date earlier than is correct.

The window depicting Jonah and the Whale (Fig.13.11) in the east central window in the corona in Canterbury Cathedral is interesting because it is of 13th century date (see Rackham, 1949). The ship shown in the window is noteworthy because of the 'myke' situated on the fore sternpost and the presence of stem bands. The furled sail and the planking with its decorative beading and nail heads are carefully and realistically depicted. The plank runs in the bow are confused, however, and the steersman is shown holding the steering oar in an unlikely manner (unless he is supposed to be "poling" the boat with an oar). This window shows clearly how ambiguity and realistic detail may be observed on one piece of iconography.

COINS

A great deal of importance has been attached to coins as a source material in ship archaeology (most recently, see Fenwick, 1978, and Muckelroy et al., 1978). The importance of coins is the result of their relative abundance and because they are often the only source of information available for certain periods and specific geographical areas (Muckelroy et al. 1978:443).

Most representations on coins are small, however, and many lack detail. The natural coin shape also tends to limit and define the way in which the ship is portrayed. Ship representations on coins may also be anachronistic, as is the 16th century ship depicted on the 1967 British halfpenny shown here (Fig.13.12). Similar anachronisms may have occured on ancient coins and it would be wrong to assume ships depicted on coins represent vessels contemporary with the minting of the coin. The gradual debasement of certain designs is an important factor in the interpretation of coin motifs.

SEALS

Many of the problems associated with the interpretation of coins may apply to seals. Seals do, however, tend to have a larger format, which favours a more realistic and more accurate representation of ship subjects. As Crumlin-Pederson has pointed out (1972:182), town seals of 13th and 14th century date are often of high quality, and towns " seem to have striven to provide the best possible pictorial record of the latest line in contemporary ships, the foundation of their urban wealth".

The factors influencing design and manufacture, and the conditions under which the seals have been preserved, generally make them a more valuable source material than coins.

TOKENS

Tokens are a little-used source, the material available being small in quantity. As a source material, tokens share many of the problems of interpretation usually associated with coins. They can be classified separately because of their different function and because of their diversity.

The example shown here is a rubbing of a lead "monk's token" found in Broad Street, Ely on 1968 (now on display in Ely Museum). These tokens were struck each year to celebrate the feast of St. Nicholas, when a singing boy from the church community would be elected to act as bishop for one week. The boy bishop would issue his own tokens to commemorate this event.[1] The place and date of manufacture is unknown but it is possible this example may be of 13th or 14th century date. As can be seen (Fig.13.13), the ship depicted on this token is very stylised. The arching curved hull appears to be dictated by the token's shape but the top, the fore and after castles and plank scarfs are clearly shown.

MURALS

Ships are not uncommon in medieval murals and many European examples have been quoted and discussed by ship archaeologists. It is almost impossible to generalise about the value of murals as source material, some murals are realistic and detailed, such as the example from Skamstrup Church, Denmark (Crumlin-Pedersen, 1965:132), others are more concerned with narrative, decorative or artistic effect, such as the mural in Kirke Hyilinge Kirke, Denmark (Liebgott, 1973:147-8) where the ship is made almost symmetrical to fit the available space.

The example illustrated here (Fig.13.14) comes from Breage Church in Cornwall. The small church lies four miles to the west of the town of Helston, overlooking Mounts Bay, an important shelterage for shipping. The present church was built in 1460 and several medieval murals were discovered during restoration work in 1890-91.

The ship shown here forms part of a mural depicting St. Christopher; the mural is generally dated to the 15th century by the style of clothing St. Christopher is wearing. Dating remains a difficulty, however, especially as this mural may have been retouched at a later date. As can be seen, the ship is very faded now and probably not as clear as when Brindley saw it (Brindley, 1916).

Despite some obvious clumsiness, the painting does give one the impression that the artist had, at least, been a casual observer of shipping (a safe assumption given the proximity to the coast) and it is unlikely the ship was copied from another source. The rigging is very simplified and also rather confused, but the eyes on the foot of the sail suggest that a lower bonnet was employed in addition to the one already shown in use. It is still just possible to make out the top and topsail noted by Brindley but the spar which he interpreted as either a derrick, mizzen or sprit has now almost vanished completely. The plank-runs and sternpost rudder are poorly executed but it should be remembered that the ship is peripheral to the theme of the mural. Even this simple little painting does throw some light on medieval sail plans and ship construction and helps expand our knowledge in areas where primary material is, so far, uninformative.

Fig. 13.14 St. Christopher wall mural, from Breage church, Cornwall

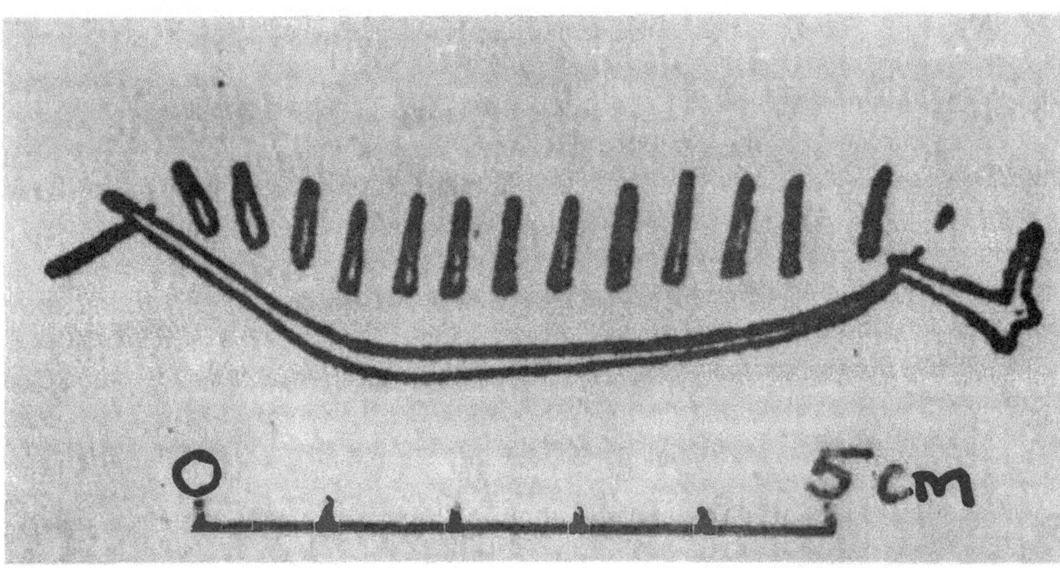

Fig. 13.15 Ship motif, from an early 7th century, Anglo-Saxon pot (after Myres and Green)

METALWORK

Ships depicted in metalwork may take several forms, from those engraved in metal (such as that on the prehistoric sword blade from Rørby, Kalundborg, Denmark) to the iron applique of ships on the church doors at Staplehurst in Kent and Stillingfleet in Yorkshire.

Although limited in design by its function and limited also by its state of preservation, the Staplehurst example does suggest something about the form and shape of Anglo-Saxon hull types (see Fenwick, 1978:Fig.8.9a and p.208). It is quite likely that many other ships depicted in metalwork still await recognition, for it was obviously a popular medium for decoration in the medieval period.

PORTABLE OBJECTS

The variety of portable objects bearing ship representations is surprisingly large and includes weather-vanes, rune-sticks, gaming pieces, jewellery and weapons. Pottery is another source of information, the Greek figured pottery being much quoted (see for instance, Casson, 1973. and Humphreys, 1978). These motifs may be very detailed or merely crudely executed ship designs, such as the one of early 7th century date illustrated by Myres and Green in their book on the Anglo-Saxon cemeteries at Caister-by-Norwich (1973:118). The ship is very stylised and not easy to interpret, the strake on the bow may be a gangplank, a steering oar (paralleling that on the Blussus carving discussed by Ellmers, 1978:4) or a feature associated with some story from Germanic legend with which this motif is associated (Fig.13.15).

CONCLUSION

Because of the diverse nature of iconographic source material, it is important to consider each example individually. The material is of undoubted value to ship-archaeology but its value will only be realised if the material is recorded and discussed in its entirety. Only rarely is the accurate depiction of a contemporary ship design the primary and sole aim of the ancient craftsman.

Note

1. I am grateful to Mr. Holmes of Ely for this information.

REFERENCES

Bonino, M., 1978. "Lateen-rigged medieval ships. New evidence from the wrecks in the Po Delta (Italy) and notes on pictorial and other documents." Int. Jour. Naut. Arch, 7:9-28.

Bowen, E. G., 1972. Britain and the Western Seaways. London.

Brindley, H. H., 1916. "Medieval and sixteenth century ships in English churches." from the Cambridge Antiquarian Society's Communications. XVII. Cambridge.

Brøgger, A. W., and Shetelig, H., 1971, The Viking Ships, their ancestry and evolution. London.

Casson, L., 1973. Ships and Seamanship in the Ancient World. Princeton.

Crumlin-Pedersen, O., 1965. "Cog—Kogge—Kaag." Aarbog Handåls og Søfartsmuseet pa Kronborg. 81-144.

- 1972. "The Vikings and the Hanseatic merchants." in G. F. Bass (ed). A History of Seafaring based on underwater archaeology. 181-204. London.

Ellmers, D., 1969. "Keltischer Schiffbau." Jahrbuch des Romisch Germanischen Zentralsmuseums Mainz, 16 :73-122.

- 1978. "Roman shipping on the Rhine." in Taylor & Cléese (ed). Roman shipping and trade: Britain and the Rhine provinces, C.B.A. Research Report No.24:1-14.

Farrell, A. W. & Penney, S., 1975. "The Broighter Boat: a reassessment." Irish Archaeol. Research Forum, 1.2:51-61. Belfast.

Farrell, A. W., (forthcoming), "The Armagh Cross and Dunluce Castle ship representations—some problems of interpretation." Irish Archaeol. Research Forum. Belfast.

Fenwick, V. H. F. (ed.) 1978. The Graveney Boat. B.A.R. Oxford.

Greenhill, B., 1976. Archaeology of the Boat. London.

Humphreys, S. C., 1978. "Artists' Mistakes." Int. Jour. Naut, Arch., 7 :78-9.

Johnstone, P., 1964. "The Bantry Boat." Antiquity, 38:277-284.

Kloner, A. & Wachsmann, S., 1978. "A ship graffito from Khirbet Rafi (Israel). Int. Jour. Naut. Arch., 7:227-244

Liebgott, Niels-Knud. 1973. "A wooden Norwegian calender of 1457 with ship graffiti." Int. Jour. Naut. Arch., 2:147-158.

Liversidge, J., 1973. Britain in the Roman Empire. London.

McGrail, S., 1974. The Gokstad faering, pt. 1. Maritime Monographs and Reports, 11. Greenwich.

McGrail, S. & Farrell, A. W., 1979. "Rowing: aspects of the Ethnographic and Iconographic evidence." Int. Jour. Naut. Arch, 8:155-166.

Marsden, P., 1972. "Ships of the Roman Period and after in Britain." in G. Bass A History of Seafaring based on underwater archaeology. 114-132. London.

Muckelroy, K., Haselgrove, C. & Nash, D. 1978. "A pre-Roman coin from Canterbury and the ship represented on it." Proc. Prehis. Soc., 44:439-444.

Myres, J. N. L. & Green, B., 1973. The Anglo-Saxon cemeteries of Caister-by-Norwich and Markhill, Norfolk.

Olsen, O. & Crumlin-Pedersen, O., 1967. "The Skuldelev ships." Acta Archaeologica, 38:73-174.

O'Riordain, B., 1971. "Excavations in Dublin." Med. Arch, 15:73-86.

Peake, W. Brian, 1930. "Luddesdown Court." reprinted from the Kent Education Gazette. Maidstone.

Pritchard, V., 1967. English Medieval Graffiti. Cambridge.

Rackham, B., 1949. The Ancient Glass of Canterbury Cathedral. London.

Smith, D. J., 1970. "The Mosaic Pavements." in A. L. Rivet (ed.) The Roman Villa in Britain. 71-125. London.

Steer, K. A. & Bannerman, J. W. M., 1977. Late Medieval Monumental Sculpture in the West Highlands. Edinburgh.

Stenton, F. (ed.) 1957. The Bayeux Tapestry. London.

Stenton, F. 1971. Anglo-Saxon England. Oxford.

Throckmorton, P., 1972. "Romans on the sea." in G. Bass (ed.) A History of Seafaring based on underwater archaeology. 66-86. London.

Throckmorton, P. & J., 1973. "The Roman wreck at Pantano Longarini." Int. Jour. Naut. Arch, 2:243-266.

Toby, A.Steven. 1974. "Another look at the Copenhagen sarcophagus." Int. Jour. Naut. Arch, 3:205-211.

KEYWORDS

Medieval.
North West Europe.
Boats and ships.
Iconography; stone carving; graffiti; models; manuscript illumination; wood-carving; tapestries; mosaics; stained-glass; coins; seals; tokens; murals; metalwork.

14. PROTECTION AND CONSERVATION OF SHIPWRECKS

Dr. J. de Jong

Summary

With the accelerating discovery of wrecks in Lake IJssel as land reclamation continues, it has to be decided whether to preserve them *in situ* or to excavate and conserve them. The value and size of the object, the wood species and its condition are all considered during this decision-making. A method for *in situ* preservation, and two methods for conserving little-degraded wood, are described.

INTRODUCTION

During the seventeenth century a plan was made in Holland to enclose the Zuyderzee and to establish a better protection of the land against the sea. It was not until 1918 that the Dutch Government, after a large flood in 1916 and a period of food scarcity during World War I, decided to have the Zuyderzee separated from the North Sea by means of a barrier dam, and to construct 5 polders in the resulting freshwater lake, Lake IJssel.

The first polder, the Wieringermeerpolder, was drained in 1930, and the barrier dam was completed in 1932. The first polder was therefore constructed in a marine environment.

The other polders have been reclaimed from a fresh water lake. The second polder, the Noordoostpolder (North-east-polder), was drained in 1942, Oostelijk Flevoland (Eastern Flevoland) in 1957, and Zuidelijk Flevoland (Southern Flevoland) in 1968. Zuidelijk Flevoland is still under development. Figure 14.1 gives an impression of the area of reclaimed land in Lake IJssel.

During reclamation and development operations in the new polders many finds of archaeological and historical value have been excavated. They vary from remains of Neolithic settlements through shipwrecks to crashed aeroplanes dating from World War II. The shipwrecks are dated from the Middle Ages to recent times. The wrecks are found here because, besides being an important fishing area, the Zuyderzee was a centre for national and international shipping. It was however a shallow and dangerous sea and it is therefore not to be wondered at that the bottom has so many shipwrecks. Uptill now about 350 of them have been found (Fig. 14.2).

The IJsselmeerpolders Development Authority is in charge of the research in this field. The work is carried out by the Scientific Department of this Authority. The Museum of Maritime Archaeology at Ketelhaven is a subdivision of the Scientific Department.

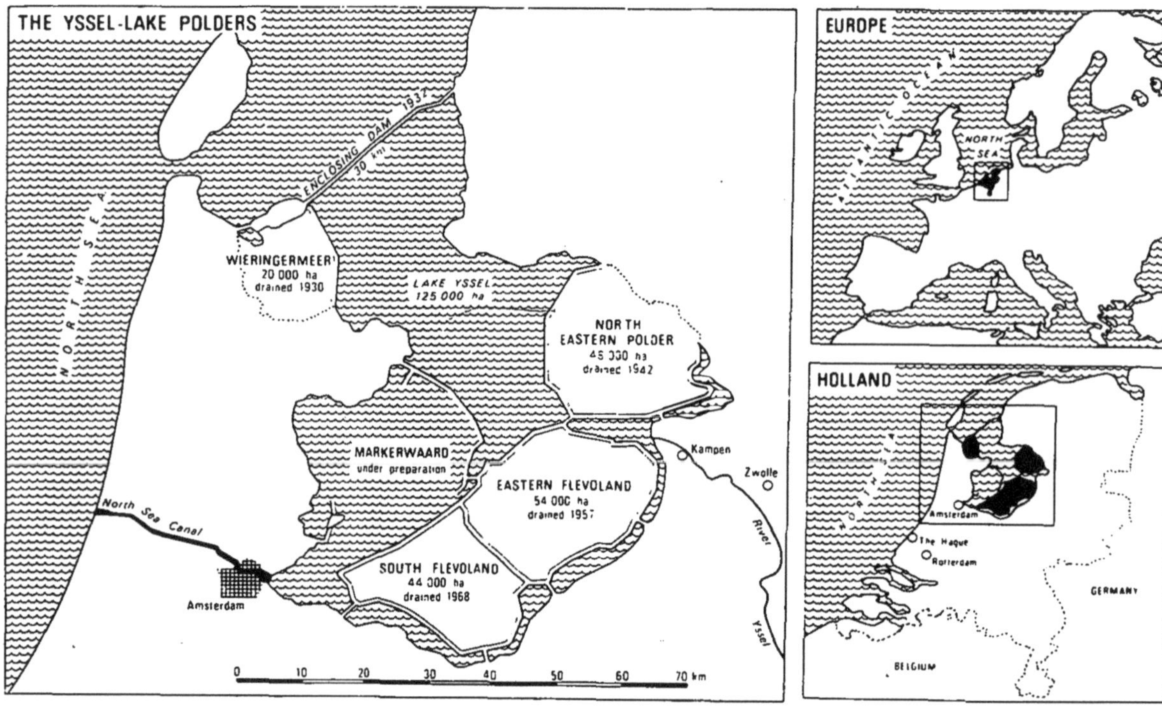

Fig. 14.1 The Lake IJssel land reclamation and development project

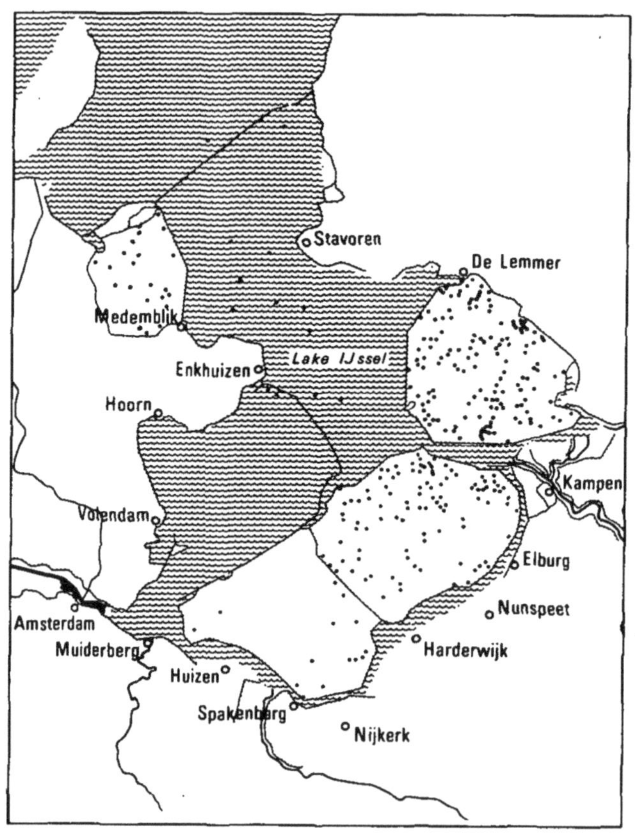

Fig. 14.2 The location of the shipwrecks found in the Lake IJssel area (1976)

A number of these shipwrecks have been excavated to study the history of the trade of shipbuilding, and life on board the ships. In several cases an entire wreck, large parts of a wreck or the ship's inventories proved to be of such interest and value that it was decided to add them to the collection of the Museum for Maritime Archaeology at Ketelhaven.

On the other hand the rate of the discovery of wrecks exceeds the excavation and research capacity. Reclamation of the new land will, due to an aeration of the upper part of the soil, result in an increased degradation of wooden objects in the aerated soil layers. So valuable objects are either to be protected *in situ* against such a degradation or they have to be excavated and to be conserved. In both cases it is essential to have information about the properties of old waterlogged wood and methods for the protection and conservation of this material.

THE PROPERTIES OF OLD WATERLOGGED WOOD

If wood is kept in a very moist or wet environment it will absorb water and eventually become waterlogged. In time changes will occur in the wood structure, and in the physical and chemical properties of the wood. These changes adversely affect the suitability of the wood for construction purposes, and are mainly caused by microbiological attack on the wood. Physical-chemical processes also contribute to breakdown.

The process of wood decay in an aquatic environment or a wet soil has been described.[1,2,3,4] The decay may occur in both aerobic and anaerobic conditions. In general the aerobic degradation of wood is considerably faster than the anaerobic degradation.

The amount of oxygen in the water saturated soil layers in which most of the shipwrecks are found in the Netherlands is small, hence disintegration of the wood has been mainly anaerobic. The process is visualised as follows.[5]

At the outset, the readily water soluble compounds in the wood will diffuse into the surrounding medium, followed by readily hydrolised compounds such as the pectins and pentosans. Then a microbiological degradation of the more stable hemicellulose and cellulose will follow. Finally mainly lignin remains, which can also be decomposed slowly by micro-organisms in an anaerobic environment. Both fungi and bacteria cause the microbiological breakdown of wood and both use extra-cellular enzymes to hydrolise hemicellulose, cellulose, and lignin. In the case of a fungal breakdown degradation spreads throughout the capillary system of the wood at least as far as the extra-cellular enzymes can penetrate into this system.3, 6, 7 The results of the attack on the wood is a considerable loss of original strength and a large increase of the shrinkage upon drying. Often the contracting capillary forces of the evaporating water exceed the strength of the capillary system of the wood and cell-collapse can be observed.

The degradation of the wood results also in a decrease of the absorbtive properties of the wood, resulting in a decreased swelling capacity of the old waterlogged wood when compared to new wood. In practice, dried and shrunken old waterlogged wood swells only a very limited extent when rewetted. The shrinkage is almost irreversible in this degraded wood. This is contrary to new wood.

The amount of shrinkage occurring in old waterlogged wood appears to be correlated with the amount of degradation of the wood as reflected in its chemical composition.

Data recorded in literature and obtained in our investigations are combined in Table 14.1 to illustrate for European oak this relation between the main chemical composition and shrinkage upon drying from waterlogged to oven-dry (105°C).

	Content in % of the oven-dry wood			Shrinkage on drying from waterlogged to oven-dry (105°C) in % of the wet dimensions		
Origin of wood	water	cellu-lose	lignin	longitu-dinal	radial	tangential
New	100-120	50-65	25-32	0.4	5	11
Utrecht Ships*	130	45	25	1	8 *	19
Zwammerdam ships *	100-120	47	23	0.5	13	21
Bremer kog	120-145	-	-	-	10	20
Bremer kog	-	46	36	-	-	-
Wasa	± 150	67	29	-	-	-
Utrecht ships	375-415	5.5	51	10	10	33
Zwammerdam ships	375-400	5.5	51	13	11	47

Table 14.1 The main chemical composition and the shrinkage of European oak heartwood at various stages of degradation
* mean values of various determinations

As can be seen from Table 14.1 a correlation between water content, lignin and cellulose content but also with shrinkage behaviour seems to be present. In order to analyse the value of this correlation we calculated for the individual data the correlation coefficients for a linear regression of the water content with the cellulose and lignin content and with the measured shrinkages.

Some of the results are shown in the Figures 14.3, 4 and 5 and summarized in Table 14.2.

Correlation of water content with:	Coefficient of correlation
cellulose content	0.84
lignin content	0.91
longitudinal shrinkage	0.94
tangential shrinkage	0.95
radial shrinkage	0.53

Table 14.2 The coefficients of correlation for a linear regression of the water content of old waterlogged European oak heartwood with the cellulose content, the lignin content and the shrinkages

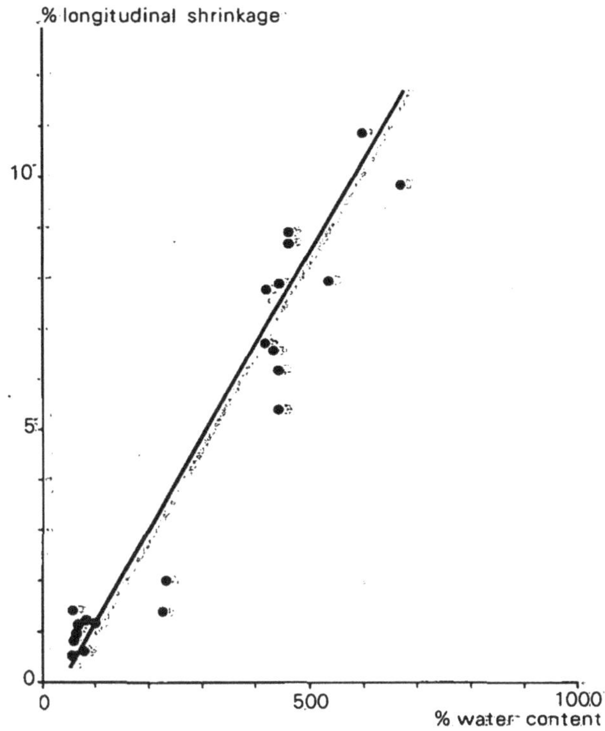

Fig. 14.3

The relation between the % longitudinal shrinkage and the % water content of old waterlogged European oak heartwood (shrinkage as a % of the wet dimensions of the wood, water content as a % of the dry wood material).

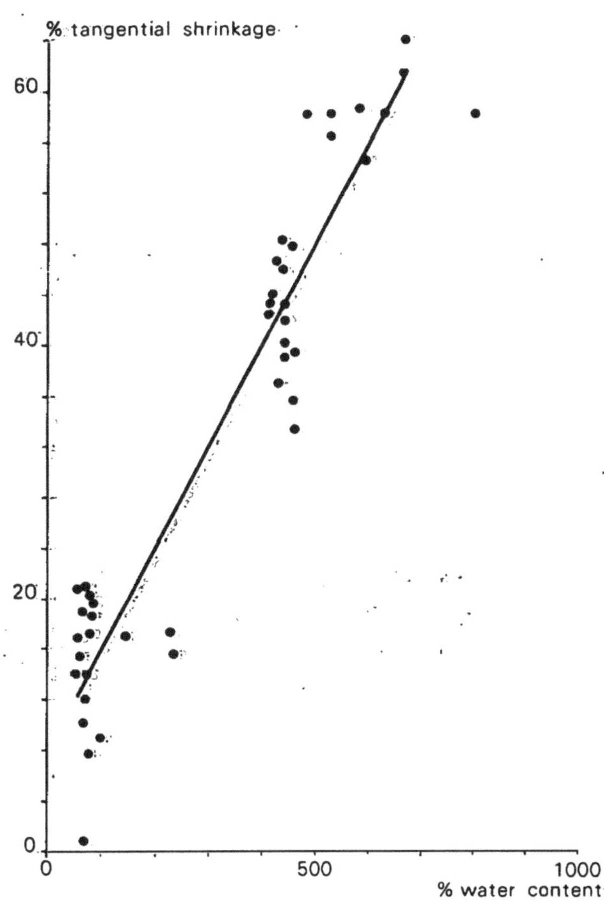

Fig. 14.4

The relation between the % tangential shrinkage and the % water content of old waterlogged European oak heartwood.

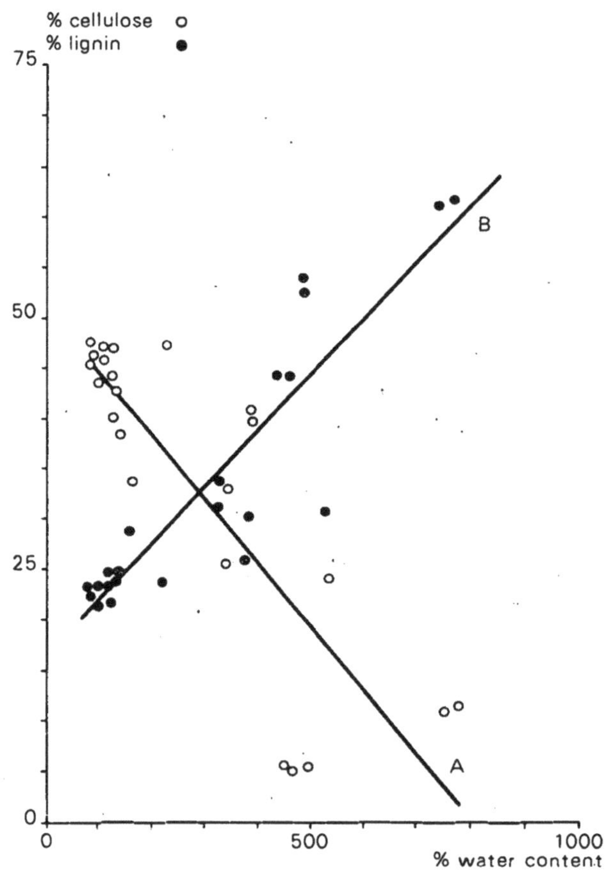

Fig. 14.5 The relation between the % cellulose content and the % water content and the % lignin content and the % water content of old waterlogged European oak heartwood (cellulose and lignin content as a % of the dry wood material)

As can be seen from Table 14.2 and the figures all correlations are significant with an exception for the radial shrinkage. However it is often difficult to measure the radial shrinkage with sufficient accuracy due to limited sample size and crack formation.

From the results shown in Table 14.2 and Figures 14.3, 14.4 and 14.5 it can be concluded that the water content of old waterlogged wood gives a satisfactory indication of the decay the wood has sustained and the chemical and physical behaviour which has to be expected. So a relatively simple test can give information about the condition of European oak heartwood. This information is essential when deciding upon the most appropriate method of conservation.

PRESERVATION OF WRECKS *IN SITU*

As indicated in the introduction a particular type of wood degradation has to be expected if the wood is found in land reclaimed from sea or lakes. During the time that these finds were covered by the sea or lake sediments the wood was attacked by anaerobic microbiological and physico-chemical processes. After the reclamation of the land, resulting mostly in a lowering of the groundwater level, the degradation of the wood will be continued by aerobic processes and hence deterioration is speeded up.[8]

When a ship was sunk in the former Zuyderzee the following happened. The wreck sinks partly in the soft sea bottom and is aided by the tidal currents and the silt transported with the river Rhine, soon covered with mud. Often parts of the wreck which remain above the level of the sea bottom are damaged by other vessels, anchors or fishing activities. Finally the wreck or its remains are entirely covered with mud. Then a polder is reclaimed and developed. At the moment of draining the soil is still muddy and completely saturated with water. In the course of any further development it is essential to remove this excess of water from the top layers of the soil. The result of this removal of water is a lowering of the soil level, called subsidence, and an increase of the amount of oxygen in the soil.

The process of subsidence is considerably, especially in the first years after drainage of a polder and then decreases gradually in time. It may amount to well over 1 m in a period of 100 years. The areated soil zone is in general limited to the upper 50-100 cm of the soil. In this area the wood of a wreck is exposed to increased degradation. These processes are illustrated in Figure 14.6.

An indication of the lifetime of wood in an aerated humid soil can be obtained from the durability classes which have been developed for new wood. This division of wood species into several durability classes is based on the lifetime of untreated heartwood, during an outdoor exposure and in contact with a humid soil in a temporate climate. European oak heartwood is, according to this division, placed on the boundary of the classes II and III. The expected lifetime will be about 15 years and for small objects even less.[4]

This increased degradation can be illustrated by the observations made on the bottom planking of a 17th century ship which was excavated in the North-eastern polder in the Lake IJssel area. This polder was drained in 1942. In

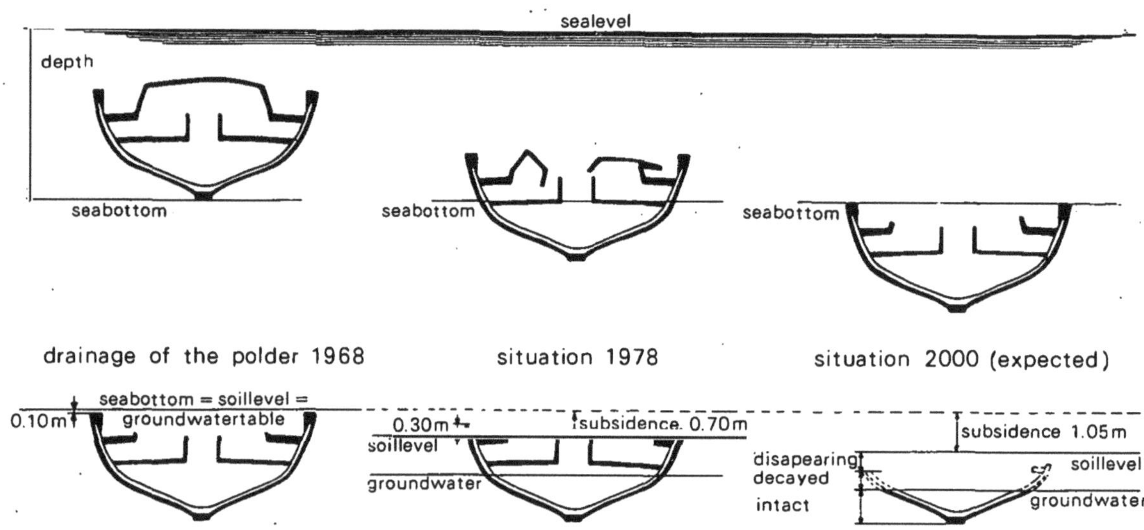

Fig. 14.6 The phenomena to which a shipwreck is exposed in the Lake IJssel area and the Lake IJssel polders

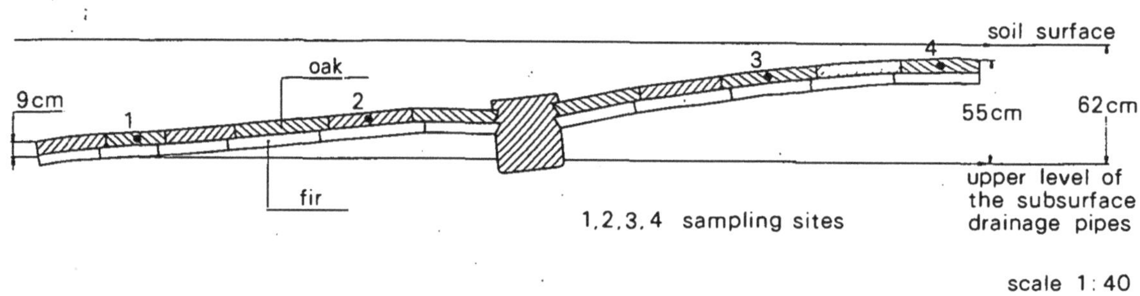

Fig. 14.7 The position of the bottom planking of a 17th century ship in relation to the level of the sub surface drainage. Sampling sites 1 and 2 (mainly) anaerobic soil, sampling sites 3 and 4 humid aerobic soil

May 1948 a ship was found at site C 48 and the wood found was described as sound and very hard.

The excavation of the wreck was delayed until 1976 and during the excavation it was observed that those parts of the wreck which had been exposed to humid and aerated soil conditions were severely damaged and decayed. So samples were taken at various sites of the bottom planking and analysed for their water, cellulose, hemicellulose and lignin content.

The sampling sites are shown in Figure 14.7; the results of the analysis are given in Table 14.3.

Sampling sites	Content in percentages of oven-dry wood			
	water	cellulose	hemicellulose	lignin
1	137	41	14	27
2	118	48	13	23
3	388	42	12	29
4	435	26	10	35

Table 14.3 The chemical composition of wood samples corresponding with the sampling sites 1, 2, 3 and 4 of Figure 14.7

From the data given in Table 14.3 it can be concluded that the wood of the samples 3 and 4 had been seriously damaged in the period between the discovery of the wreck and the actual excavation. The damage has been caused by the lowering of the groundwater table and the resulting subsidence of the soil. These effects caused the change of a slow anaerobic degradation of the wood into a fast aerobic one. From this and other simiiar observations it was concluded that it was necessary to develop a method to ensure in situ preservation of valuable objects when excavation and conservation within a short period after the discovery of the find is impossible.

The method developed is based on a local maintenance of a high ground water table and the exclusion of the access of oxygen to the object. In the clay soils of the Lake IJsselpolders this can be realised by removing the subsurface drainage pipes near the wreck and positioning a plastic sheeting vertically around and horizontally above the wreck. By giving the plastic sheeting which is above the wreck covered with a thin layer of clay a concave shape and making some holes in it rain water can be collected and used to maintain a high groundwater table at the site of the wreck. If needed additional water can be supplied.

THE CONSERVATION OF SHIPWRECKS

The available techniques for conservation of old waterlogged wood are known and reviewed in the literature.[5,8,9,10]

For the treatment of large objects impregnation with polyethyleneglycol is still the method most widely used. It has however been shown that the results of this way of conserving object depends strongly on the size of the object, the wood species and the degree of degradation of the wood.[4,8]

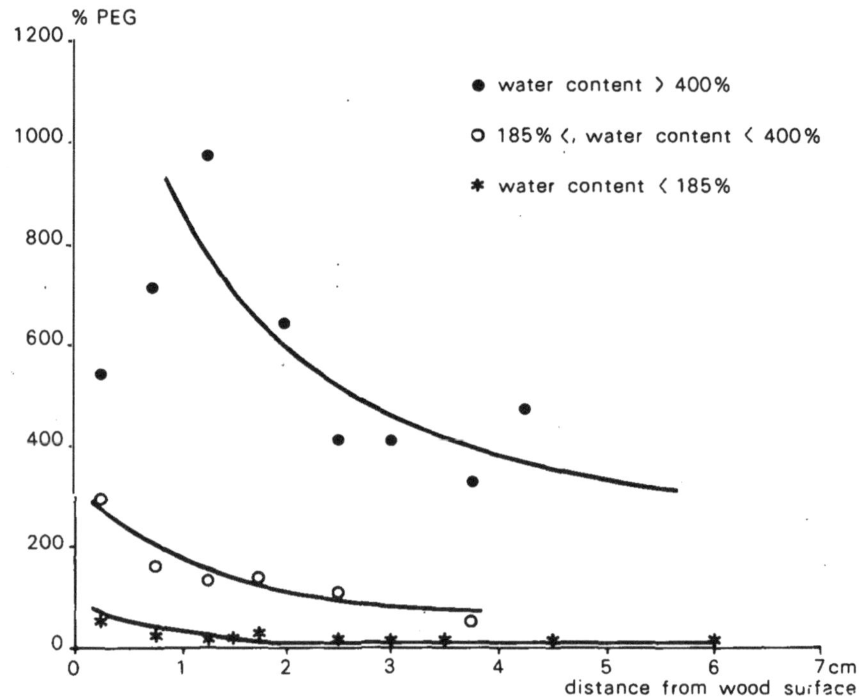

Fig. 14.8 The impregnation of PEG into European oak heartwood (waterlogged). Impregnation time approx. 2 years. Final concentration 65% PEG by weight, impregnation temperature 60°C

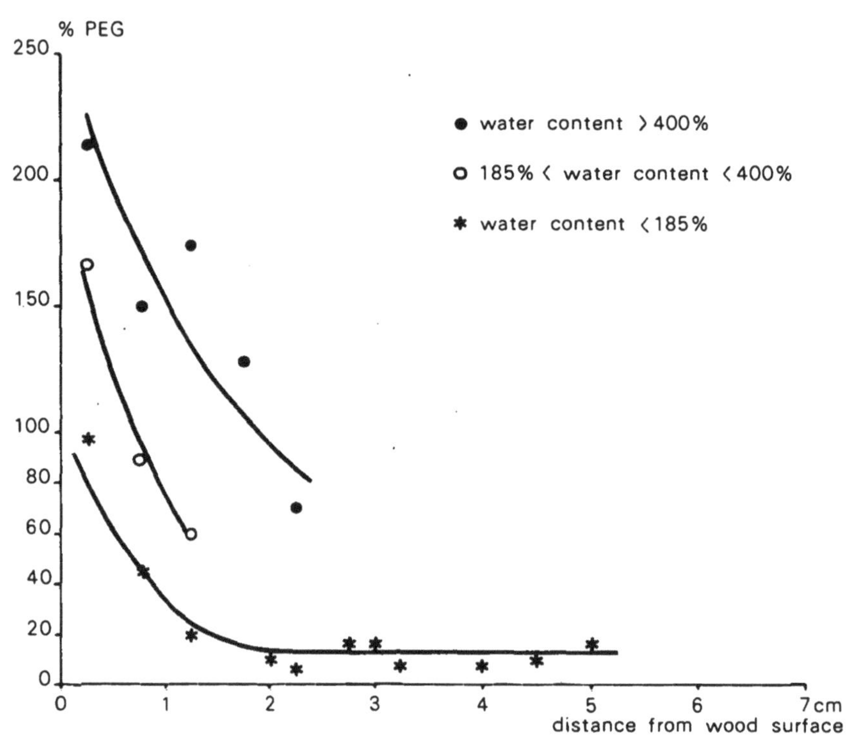

Fig. 14.9 The impregnation of PEG into waterlogged European oak after the tba-PEG method (large objects)

For the conservation of shipwrecks or large wooden objects the following methods are used at the Museum of Maritime Archaeology at Ketelhaven.

- —Impregnation with PEG 4000 at 60°C for wood having a water content over 400% (Class I wood);
- —Impregnation with PEG 4000 at 20-30°C for wood having a water content between 185-400% (Class II wood).

Both methods give good to satisfactory results.

Little; degraded wood, having a water content of less than 185% (Class III wood) cannot be treated within a feasible time as is illustrated in Figure 14.8. This figure shows the impregnation of PEG 4000 into European oak perpendicular to the longitudinal direction. The results are obtained from the centres of large objects (having a length over 3 m).

The wood was impregnated with PEG 4000 at 60°C and the concentration was raised from 0 up to 65% by weight in 18 months and then maintained constant for 6 months.

From this figure it can be seen that in the little-degraded wood insufficient PEG is present in the core to ensure a good dimensional stabilization of the wood, since the required amount of PEG is about 20-30% based on dry wood materials.[11, 12]

For this little-degraded wood new approaches for conservation are under development and being tested on a pilot plan scale. The methods are either a partial dehydration of the wood with tertiarybutyl alcohol (tba) followed by an impregnation with PEG in the alcohol-water mixture, or a controlled drying of very little decayed wood.

The tba-PEG technique has three stages. During the first stage the wood is dehydrated until azeotropic composition in the tertiary butylalcohol water mixture; then it is impregnated with PEG 4000 in this alcohol-water mixture; and at last, after the wood has been taken out of the impregnation tank and the alcohol has been allowed to evaporate, the wood is given heat treatment to ensure as much diffusion of the PEG 4000 into the cell walls as possible.

During dehydration which is done at room temperature the water is removed from the alcohol by azeotropic distillation. During this stage of the process dark coloured compounds are extracted from the wood. This results in a more natural colour of the dehydrated wood. Besides this the impregnation speed appears to be higher than in the water PEG process. It was found that the shrinkage occurring during this stage of the process is virtually independent of the dehydration speed. After dehydration the wood is impregnated with PEG 4000 dissolved in the alcohol-water azeotrope at 55°C. The concentration is raised off from 0 through 50% by volume during 10 weeks (wood samples measuring 20 x 30 x 95 cm) and is then kept constant for another 8 weeks at 55°C. Then the wood is removed from the bath and the alcohol allowed to evaporate for 2 or 3 weeks. At last the wood is given a heat treatment at 60-80°C of from a few days to two weeks.

The result of the treatment is a sound and natural looking wood that only slightly shrunk during the treatment as can be seen from Table 14.4

Stage of treatment	% shrinkage European oak tangential	radial	longitudinal	ash and elm tangential	radial	longitudinal
after dehydration	3.1	1.9	0.4	0.2	1.3	0.1
after impregnation	3.8	2.9	0.7	1.5	1.2	0.1
after evaporation of the alcohol	4.2	3.3	0.8	1.7	2.4	0.5
after heat treatment	4.5	3.2	0.8	1.9	3.2	0.6

Table 14.4 Mean shrinkage values of Class III European oak, ash and elm at different stages during the tba-PEG 4000 treatment of wood. Shrinkage as a % of the wet dimensions

As can be seen the observed shrinkages are small. Until now very satisfactory results have been obtained, even with difficult objects, containing little degraded heartwood, highly degraded heartwood and sapwood, and even bark.[4]

The impregnation results as a function of the impregnation depth (perpendicular to the longitudinal direction) are given in Figure 14.9 for classes I, II and III European oak heartwood.

The surface concentrations in classes I and II wood are lower than in the water-PEG process as a result of the lower bath concentration and the heat treatment which allows for a more even PEG distribution.

In class III wood the PEG concentration in the core of the wood is about 12%, which is less than the required levels as indicated in literature. Up till now, however, no problems arose in the treated objects during 4 years after their treatment and storage in both laboratory and the (rather damp) museum workshop.

It will be clear that the tba-PEG 4000 technique is more complicated and expensive than the water-PEG technique. Hence we looked for different, and less expensive possibilities for conservation. It was found that in some instances the wood of wrecks discovered in the soil of the Lake IJssel Polders was of such a good quality that the wreck could be dried at the excavation site without the occurrence of large shrinkages, cracking and deformation phenomena. This was observed for a 19th century wreck but also for fragments of considerably older wrecks.

This led to the conclusion that in some instances, when the wood is very little deteriorated, drying of the wood might be a possibility. To investigate this possibility in more detail samples of timber (oak and pine) were taken from a 17th century wreck that is kept under a shower installation in the museum at Ketelhaven. The samples (approx. 200 x 10 x 25 cm and 200 x 4 x 20 cm) were dried under laboratory conditions (20°C, 40-60% r.h.) and the shrinkages were measured. It was found that the observed shrinkages were comparable to those occurring in new wood. During the drying of the wood only a very slight surface cracking occurred.

In 1978 samples have also been taken from an 18th century wreck found in a marine environment. This wood was also very little degraded and it was decided to measure the occurring shrinkages upon drying. The observed shrinkage data are given in Table 14.5.

	% shrinkage		
	tangential	radial	longitudinal
new European oak heartwood	10.6	4.8	0.4
Lake IJssel wreck oak heartwood	12.1	-	0.4
marine environment wreck, oak heartwood	5.8	4.2	1.1
new pine	7.7	3.9	0.4
Lake IJssel wreck pine	6.3	-	0.4

Table 14.5 The mean shrinkage values for new and old waterlogged wood. % shrinkage based on the wet dimensions of the wood. Drying from waterlogged to oven-dry (105°C).

These observations, combined with the collected evidence on the possibilities of drying complete wrecks or large segments of ships, encouraged the start of preparations for controlled drying of two wrecks in the museum at Ketelhaven. One wreck (approx. 12m long) is being dried under controlled conditions, a second and larger one (25 m long) will follow.

REFERENCES

1. Barkman, K., 1965. The preservation of the Wasa. Wasastudier 5, Statens Sjöhistoriska Museum, Stockholm.
2. Boutelje, L. B. and Kiesling, H., 1964. On water-stored oak timber and its decay by fungi and bateria. Archiv. Mikrobiol. 49, (3), 305-314.
3. Levy, J. F., 1977. Degradation of wood. In: Sources and techniques in boat archaeology. BAR Supplementary Series 29, pp. 15-22.
4. Jong, J. de, 1977. Conservation of old waterlogged wood. In: Sources and techniques in boat archaeology. BAR Supplementary Series 29, pp. 23-44.
5. Mühlethaler, B., 1973. Conservation of waterlogged wood and leather. ICOM traveaux et publication XI, Ed. Eyrolles, Paris.
6. Stamm, A. J., 1964. Wood and cellulose Science. Ronal Press Company, New York.
7. Hashimoto, Y. and Nisizawa, K., 1963. Advances in enzymic hydrolysis of cellulose and related materials, ed. E. T. Reese, Pergamon Press, London.
8. Jong, J. de, 1977. The conservation of old waterlogged wood from shipwrecks found in the Netherlands. In: Biodeterioration Investigation Techniques Interscience.

9. Oddy, W. A. ed., 1975. Problems in the conservation of waterlogged wood. Maritime monographs and Reports, no. 6.

10. Brorson Christensen, B., 1970. The conservation of waterlogged wood in the National Museum of Denmark, Copenhagen.

11. Stamm, A. J., 1959. Effect of polyethyleneglycol on the dimensional stability of wood. Forest Prod. J. 9, 375-381.

12. Ambrose, W., 1972. The treatment of swamp degraded wood by freeze-drying. Report to the ICOM committee for conservation. Madrid.

KEYWORDS

Medieval and later.

Zuiderzee; Netherlands.

Waterlogged wood.

Conservation; preservation; wood degratadion; wood shrinkage; oak heartwood; impregnation; dehydration.

www.ingramcontent.com/pod-product-compliance
Lightning Source LLC
LaVergne TN
LVHW071205210926
848482LV00015B/728
9780860540687